T0317722

How to Commercialize Chemical Technologies for a Sustainable Future

How to Commercialize Chemical Technologies for a Sustainable Future

Edited by
Timothy J. Clark and Andrew S. Pasternak
GreenCentre Canada
Kingston
Ontario
Canada

This edition first published 2021
© 2021 by John Wiley & Sons Ltd

The right of Timothy J. Clark and Andrew S. Pasternak to be identified as the authors of the editorial material of this work has been asserted in accordance with law.

Registered Offices
John Wiley & Sons, Inc., 111 River Street, Hoboken, NJ 07030, USA
John Wiley & Sons Ltd, The Atrium, Southern Gate, Chichester, West Sussex, PO19 8SQ, UK

Editorial Office
The Atrium, Southern Gate, Chichester, West Sussex, PO19 8SQ, UK

For details of our global editorial offices, customer services, and more information about Wiley products visit us at www.wiley.com.

Wiley also publishes its books in a variety of electronic formats and by print-on-demand. Some content that appears in standard print versions of this book may not be available in other formats.

Limit of Liability/Disclaimer of Warranty
In view of ongoing research, equipment modifications, changes in governmental regulations, and the constant flow of information relating to the use of experimental reagents, equipment, and devices, the reader is urged to review and evaluate the information provided in the package insert or instructions for each chemical, piece of equipment, reagent, or device for, among other things, any changes in the instructions or indication of usage and for added warnings and precautions. While the publisher and authors have used their best efforts in preparing this book, they make no representations or warranties with respect to the accuracy or completeness of the contents of this work and specifically disclaim all warranties, including without limitation any implied warranties of merchantability or fitness for a particular purpose. No warranty may be created or extended by sales representatives, written sales materials or promotional statements for this work. This work is sold with the understanding that the publisher is not engaged in rendering professional services. The advice and strategies contained herein may not be suitable for your situation. You should consult with a professional where appropriate. The publisher, editors, and authors shall not be liable for any loss of profit or any other commercial damages, including but not limited to special, incidental, consequential, or other damages. The fact that an organization, website, or product is referred to in this work as a citation and/or potential source of further information does not mean that the publisher and authors endorse the information or services the organization, website, or product may provide or recommendations it may make. Further, readers should be aware that websites listed in this work may have changed or disappeared between when this work was written and when it is read. The publisher, editors, and authors shall not be liable for any loss of profit or any other commercial damages, including but not limited to special, incidental, consequential, or other damages.

For general information on our other products and services or for technical support, please contact our Customer Care Department within the United States at (800) 762-2974, outside the United States at (317) 572-3993 or fax (317) 572-4002.

Wiley also publishes its books in a variety of electronic formats. Some content that appears in print may not be available in electronic formats. For more information about Wiley products, visit our web site at ww.wiley.com.

Library of Congress Cataloging-in-Publication Data applied for
HB:ISBN: 978-1-119-60484-6

Cover Design: Wiley
Cover Image: © cherezoff/Shutterstock

Set in 9.5/12.5pt STIXTwoText by Straive, Chennai, India
Printed and bound by CPI Group (UK) Ltd, Croydon, CR0 4YY

C9781119604846_180521

I would like to dedicate this book to my father, Jack Pasternak, who was an active chemical technology entrepreneur in his own right. He was a great supporter of the creation of this book, but unfortunately passed away before its publication. I would also like to thank the ongoing support and encouragement of my wife Maxine and son David who put up with my spending many evening hours on this project.

Andrew S. Pasternak

I also dedicate this book to my father, Ron Clark (1945–2012), who was a chemical engineer and an incredibly supportive and loving dad. He was a meticulous technical editor and would have been very proud of this book. I want to acknowledge my wife Lauren who is a constant source of loving encouragement and patience, and my two sons, Calum and Ciaran, who are the most fun people I know. Lastly, I am very grateful to my mother Sylvia, brother Andrew, and sister-in-law Merrylee for their love and support.

Timothy J. Clark

Contents

List of Contributors

Harish Bhaskaran
Dept. of Materials
University of Oxford
Oxford
UK

Jason Clark
Braskem America
Cambridge, MA
USA

Timothy J. Clark
GreenCentre Canada
Kingston, ON
Canada

Matthew Cohen
Pangaea Ventures
Phoenix, AZ
USA

Janine Elliott
Los Angeles Cleantech Incubator
Los Angeles, CA
USA

Andrew Ellis
NORAM Engineering and BC Research
Vancouver, BC
Canada

Richard E. Engler
Bergeson & Campbell, P.C. and The Acta
Group
Washington, DC
USA

Tess Fennelly
GreenSustains, Inc.
Minneapolis, MN
USA

Gert-Jan Gruter
Avantium
NV Amsterdam
The Netherlands

Lauren Heine
ChemFORWARD
Spokane, WA
USA

Peiman Hosseini
Bodle Technologies
Oxford
UK

Shawn Jones
White Dog Labs Inc.
New Castle, DE
USA

James Lockhart
NORAM Engineering and BC Research
Vancouver, BC
Canada

Kira Matus
Division of Public Policy
HKUST, Clearwater Bay
Kowloon
Hong Kong

Sean Monkman
CarbonCure Technologies Inc.
Dartmouth, NS
Canada

Aidan R. Mouat
Hazel Technologies
Chicago, IL
USA

Andrew S. Pasternak
GreenCentre Canada
Kingston, ON
Canada

Rohit Sood
Spinverse Oy
Espoo
Finland

Nick Sutcliffe
Mewburn Ellis LLP
Cambridge
UK

Tom van Aken
Avantium
NV Amsterdam
The Netherlands

Jennifer Wagner
CarbonCure Technologies Inc.
Dartmouth, NS
Canada

Andrew White
CHAR Technologies Ltd.
Toronto, ON
Canada

Margaret H. Whittaker
ToxServices LLC
Washington, DC
USA

1

Introduction
Timothy J. Clark and Andrew S. Pasternak

GreenCentre Canada, Kingston, ON, Canada

1.1   What Is This Book About?

The fundamental impact of the chemical industry is pervasive throughout the world. Every product, material, and object we own or use owes its existence in some way to this vital sector. Our food supply, medicines, clothing, and mobile devices all depend on chemistry. Even products or services that upon first glance do not obviously involve chemistry undoubtedly do for some secondary purpose such as their storage, transportation, or delivery [1]. In addition, the chemical industry plays a dominant role in the global economy, being responsible for more than $5 trillion in revenue and 20 million jobs worldwide [2]. It has broadly contributed to our technological progress over the last 200 years.

The industry has also brought problems that are increasingly recognized as "must solve" to ensure long-term human health along with environmental, economic, and even geopolitical stability. Examples include the intrinsic safety of chemicals available in the marketplace, hazardous materials released into the environment during manufacture, and the materials' use and disposal, all of which are receiving more attention than ever before [3]. This also unsurprisingly coincides with the increasing number of peer-reviewed, data-driven reports demonstrating negative long-term effects on the planet and its inhabitants [4]. Climate change and masses of nonrecyclable plastics littering the ocean are just two obvious examples.

These challenges are daunting but not insurmountable, especially as there is no shortage of technical innovations and advances in sustainable chemistry emerging from around the world. The academic community is constantly discovering promising new chemical technologies, and more importantly entrepreneurs are becoming empowered and encouraged to *bring them to market*. This is imperative as any anticipated or quantified sustainable benefit associated with a given technology will never realize its potential while it remains a laboratory-scale research project. In other words, it is *only through the*

How to Commercialize Chemical Technologies for a Sustainable Future, First Edition.
Edited by Timothy J. Clark and Andrew S. Pasternak.
© 2021 John Wiley & Sons Ltd. Published 2021 by John Wiley & Sons Ltd.

development, scale-up, and commercial deployment of sustainable chemical technologies that these challenges can be overcome. The issue at hand is how to advance a promising technology down the development path to the point where it has been validated, demonstrated to be economically competitive, and scaled to meet customer demand. Unlike large multinationals with sizable resources to address commercialization challenges, the entrepreneur developing a sustainable chemical technology is severely resource limited and faces significant barriers. When it comes to raising the required funds, they are often trapped in a catch-22 situation: funding requires validation and scale, while validation and scale require funding.

It is the entrepreneur who will undoubtedly play a crucial role in deploying the required technologies that ensure we maintain our quality of life while not robbing future generations of the same [5]. This is the essence of sustainability. Multinational companies will continue to invest and innovate, but their resources are not infinite, and "out-of-the-box" thinking and nonincremental solutions often pose a challenge to bureaucratic and conservative corporations that must answer to their shareholders [6]. Many large companies today recognize this position and are looking to support and partner with start-ups developing attractive technologies. One could argue that the future success of larger companies is at least in part dependent on the success of these entrepreneurs.

The environmental challenges associated with the chemical industry can be met by providing the budding entrepreneur with the training and skills needed to commercialize a sustainable chemistry technology. There is a significant knowledge gap between how to conceive and test an innovation and how to actually get it to market. We created this book to help address this gap. The skills required to create, operate, and grow a company are generally not part of the curriculum in current chemistry or engineering programs. While elements may be taught in more progressive departments, it is certainly not in any comprehensive manner. Relevant courses and training programs for the budding entrepreneur are becoming increasingly available, but these are not chemistry-specific and may be deemed a distraction to the student focused on their research projects. In addition, there is often trepidation on the part of chemists to take advantage of these offerings as they are often far removed from their past experiences.

This book will describe the steps, decision points, and hurdles faced by innovators developing sustainable chemical technologies and offer practical tactics and strategies for confronting them. This includes aspects of product/process development, scale-up, market landscape analysis, regulatory frameworks, strategic partnering, intellectual property management, and financing.

To the best of our knowledge, there is currently no other book on the market that addresses this broad topic. Many texts have been published about the general commercialization of technologies [7]. However, few target the chemical innovator, and none is specific to the commercial deployment of sustainable technologies. One of our overarching goals in preparing this book was not to create a comprehensive, lengthy tome that will just sit on your shelf. Instead, our intention was to offer a relatively concise guide that includes practical advice as you consider taking the entrepreneurial path.

Overall, the purpose of this book is to provide the following:

- Awareness and information on the many steps required to commercialize a sustainable chemical technology
- Guidance for making appropriate strategic choices when creating and subsequently growing a new venture
- Motivation and inspiration via success stories of early-stage companies that have been effectively passing through the various stages of technology commercialization

1.2 What Is a Sustainable Chemical Technology?

It is important to establish how we have chosen to define a "sustainable" chemical technology. Definitions abound, and controversies have arisen over linguistic nuances [8]. One basic definition (described in Chapter 3) is provided by the Organisation for Economic Co-operation and Development (OECD), which defines sustainable chemistry as "a scientific concept that seeks to improve the efficiency with which natural resources are used to meet human needs for chemical products and services." However, trade associations, individual companies, governments, and many nongovernment organizations all have variously differing definitions – many in ways that (not surprisingly) lend credence to their own mandate or beliefs. The term "green" also has numerous definitions and is applied, sometimes incorrectly, synonymously [9, 10].

For the purpose of this book, we will use a relatively simple definition: *a sustainable chemical technology offers a demonstrated environmental benefit(s) while remaining economically competitive with existing technologies.* This is a broad definition as it addresses three key elements.

The first is most obviously the *demonstrated environmental benefit*. A chemical technology can be deemed "sustainable" if it benefits at least one aspect of the environment. Examples include protecting the environment by using technologies that improve water-use efficiency and treatment and reducing waste material production and release. These benefits can also present themselves in technologies that are not intrinsically environmental in nature but serve a greater purpose of reducing energy or resource use and thus, on balance, will improve the environment. A base metal catalyst that can replace stoichiometric reagents and lower the energy input required for a given process or bio-derived plastics that can be controllably degraded and recycled are examples. Comprehensive quantification of various metrics and subsequent life-cycle analyses – topics not extensively covered in this book but available from multiple sources [11, 12] – are required prior to making any formal claim regarding environmental benefits. The results of these analyses can be surprising. There are instances where technologies may on the surface appear to be beneficial for the environment when in fact it is later proven otherwise [13]. The converse can also be true.

The aspect of *economic competitiveness* may be considered by some to be less worthy and should therefore not be included in the definition. We wholeheartedly disagree. A technology that is not cost-competitive in its respective market will simply not be adopted regardless of any environmental benefit. Government regulations or subsidies can offer short-term economic attractiveness, but it is a risky proposition to base a business on the whims of a governing party. Even a panacea chemical technology won't have its desired effect on the environment if it is never sold to, or used by, someone. Furthermore, the so-called "green premium," the additional cost of an environmentally friendly product, often negatively affects the consumer buying decision. Although many consumers claim to be eco-conscious, in practice a majority are reluctant to pay more for environmentally superior products [14].

Finally, *existing technologies* must be taken into account when assessing the sustainability of an innovation. No technology is evaluated in a vacuum. Everything is relative to what has already been developed and being used by customers today. You cannot make a claim about any sustainable benefits without comparing them to a baseline of existing technologies. And there *are always existing technologies*. In our experience, one must be skeptical of claims of a completely novel innovation that will revolutionize the industry. Perhaps it will, but close comparison to existing technologies often paints a different picture. This concept applies to the two aspects described earlier; both the environmental benefits and economic competitiveness must be evaluated relative to existing technologies.

1.3 Commercializing Sustainable Chemical Technologies Is Challenging

Chemical technologies are particularly difficult to commercialize regardless of your level of experience or motivation, as they face a complex set of challenges impacting time to market. The challenges include the following:

- Long development cycles
- Sheer size and relative inertia of the chemical industry and supply chains
- Environmental health and safety concerns
- Regulatory climate and approval processes
- Large amount of capital and financing required

Early-stage companies are often cash-poor, which limits their access to talent, specialized lab infrastructure, and expensive instrumentation and capital equipment required to scale their technology. Funding is always an issue, given that early-stage technologies do not align well with the criteria of the investment community. Many investors are wary of technologies that have not been scaled and validated in an industrial setting (providing a clearer cost profile), while entrepreneurs are reluctant to concede high amounts of equity on what they believe to be low valuations. This mismatch of intentions often leads to lost opportunities with undeveloped technologies and companies never realizing their potential.

One can further appreciate the challenges of commercializing a sustainable chemical technology by comparing to other sectors. As an example, consider the case of commercializing new application software. The intention is not to diminish the many challenges of software commercialization (and there are many), but to demonstrate the difference and scale of the challenges that chemical technologies face. See the following comparative table:

	Application software technology	Sustainable chemical technology
Overall time to market	6–18 months	5–10 years
Development stage		
Operating space required	Commercial or residential office	Appropriate chemical laboratory infrastructure to safely handle reagents and materials with varying hazards
Equipment needed	Computer(s)	Process reactors and ancillary equipment, characterization and analytical instrumentation
Cost of initial equipment requirements	<$5000	>$1 million
Investment required	Low	High
Health and safety requirements	None	High: care is required handling chemicals and equipment
Scale-up stage		
Additional skills required	Minimal	Process engineering and scale-up expertise
Cost	<$200 000	$5–10 million
Strategic partner investment required	Low: test sites can evaluate software at minimal cost	High: partner needs to invest resources to pilot the technology
Health and safety requirements	None	Very high: piloting at larger scales brings increased safety risks
Manufacturing Stage		
Additional equipment requirements	Practically none	Extensive: commercial-scale reactors or production equipment required
Distribution costs	Minimal	High
Regulatory requirements	Minimal	Very high
Health and safety requirements	None	Very high: any health and safety incidents can be disastrous

What is readily apparent with these two cases is the significant amount of resources and time required to advance a chemical technology to market. The risks are also significantly higher as it costs a lot more to fail commercializing a chemical technology than a software application. This translates directly to higher cost of capital and funding challenges.

1.4 Who Should Read This Book?

This book is targeted at innovators who are considering creating, or entrepreneurs in the early stages of running, a sustainable chemistry start-up company. In our experience, entrepreneurs in this sector often come from one of two backgrounds. Although both groups bring a strong set of skills, they are often incomplete. This book was written to begin the process of raising awareness and providing problem-solving tactics so that you can better navigate the many challenges you will encounter. This knowledge will position you to make smarter, more strategic decisions as you move forward with your business more efficiently.

The first group are undergraduate/graduate students or post-doctoral researchers who want to commercialize a technology on which they were working during their studies. They may be wondering, "What exactly needs to be done to transition my discovery into a successful business?" We have observed a greater number of chemistry graduates and professors who have invested considerable time and energy into their research projects and now desire to take them to the next level. These individuals have become highly trained and motivated technical experts but have little understanding of the next steps beyond their lab bench. The skills needed are not taught in chemistry departments and are difficult to obtain in a short time frame. Courses on entrepreneurship and mentorship resources are available, from MBA-type classes to training programs offered by university incubators or accelerators. However, these educational resources are often not tailored to the unique challenges of sustainable chemical technology commercialization.

The second group involves more experienced entrepreneurs from other sectors (not the chemical industry) who now want to make a difference by commercializing a sustainable chemical technology. These individuals were often successful with their previous ventures and have greater amounts of capital and credibility at their disposal. They have business acumen but a limited understanding of chemistry and the nuances of the commercial pathway in the chemical industry. Although passionate and accomplished, they often underestimate the considerable challenges, resources, and skill sets required to advance a chemical technology compared to their previous entrepreneurial work in a different sector.

1.5 Structure of This Book

This book is divided into four distinct parts.

The first part, "Laying the Foundation," focuses on the foundational elements of a sustainable chemistry start-up company. Chapter 2, "Marketing and Landscape Analysis," explains that successfully marketing and commercializing a sustainable chemical technology requires the same market-driven discipline that is needed to commercialize any other new technology. As such, it is critical to understand what customers need, who the competition is, and how you can successfully differentiate your product to those customers. The chapter describes an "application framework" for the creation of an effective and realistic marketing plan. It will provide templates and tools to interview your potential clients and methods to uncover the information that is most important in developing a true marketing-based (as opposed to sales-based) approach.

Chapter 3, "Determining the True Value of a Sustainable Chemical Technology," explains that a product's market value and "true" value can often be different when sustainability factors are taken into account. Starting with the 17 United Nations Sustainability Goals, this chapter explores various frameworks to better define and understand the true value of your innovation and how to communicate it to potential investors and customers.

Chapter 4, "Intellectual Property Management and Strategy," covers fundamental considerations when commercializing any type of invention. The chapter provides an "Intellectual Property 101" with a focus on patents but also touching on know-how, trade secrets, copyright, and trademarks. The patent application process will be described practically from idea to issued patent, with particular attention given to what is required of you at each step of the process. An honest overview is also provided of the benefits and detriments of intellectual property (IP) for the entrepreneur. Critically, some of the strategic considerations in developing an IP position (a "ring fence") will be described to protect your company's most valuable assets, along with how to create an IP package that will be attractive to potential investors.

The second part, "Political and Environmental Considerations," examines the political, regulatory, and public opinion drivers affecting the commercialization of sustainable chemistry technologies. It starts with Chapter 5, "Navigating and Leveraging Government Entrepreneurial Ecosystems for Support," which explores the array of resources available to you involving government, not-for profit, and for-profit organizations at regional or national scales in different parts of the world. These ecosystems were created to help you advance your company, so it is important to understand and leverage them as much as possible. These can involve financial assistance, expertise, lab space, and other resources.

Chapter 6, "Factoring in Public Policy and Perception," explains how public policy and perception must be understood when commercializing a sustainable chemistry technology. Local and regional governments create laws, while public perception creates buying trends, both of which can significantly impact the success of your business. A concise overview of several regions' regulatory frameworks will help ensure you can generate revenue from your technology without violating these regulations. This is followed by a discussion on how public opinion and attitudes toward environmental concerns affect consumer choices and can even dictate policy. This is especially true in chemistry and materials-based products where scientific information is often not well understood by the public.

Chapter 7, "Pre-market Approval of Chemical Substances: How New Chemical Products Are Regulated," examines a related but distinctly different topic that is often overlooked and poorly understood by the chemical innovator. The chapter addresses how new chemical products are regulated (if at all), research and development exemptions, and the importance of incorporating pre-market approval into your company's global commercialization strategy. Generic requirements for chemicals, as well as special requirements for specific products including cosmetics, drugs, and food contact substances, will be outlined.

The third part, "Springing into Action," covers topics that are central to developing and growing your start-up company in a rational manner. It begins with Chapter 8, "Navigating Supply Chains." Successful entry to any market is driven by the dynamics of the supply chain and one's ability to navigate the hurdles and create opportunities. Sustainable chemical technologies face additional barriers to adoption as supply chains are often broken or slow. Practical tools to map and understand your supply chain will be presented. This will be followed by methods on how to identify the biggest barriers to your new technology's adoption and strategies for overcoming them.

Chapter 9, "Strategic Partnering," presents the potential advantages and disadvantages of partnering from the perspectives of both the start-up and the larger industrial partner along with strategies to mitigate the risks within a partnership. Evaluating, establishing, executing, and finally closing a strategic partnership will be covered. The chapter will conclude with illustrative case studies to showcase examples of effective partnering.

Chapter 10, "Bridging the Gap 1: From Eureka Moment to Validation," and Chapter 11, "Bridging the Gap 2: From Validation to Pilot Scale-Up," are dedicated to describing how you can practically advance your technology from conception to commercial demonstration. Chapter 10 focuses on core technology development activities from the initial innovation in the lab to proving the concept to the point where seed financing is possible. At each stage, the overall team characteristics and required activities of your company will change. The chapter will also discuss how to efficiently de-risk the technology by identifying bottlenecks so that your product's time to market can be accelerated.

Chapter 11 addresses the many aspects of scaling up your technology to meet market needs. Due to the breadth of these topics, it is divided into two parts. The first discusses required activities before any pilot unit is actually built. These activities include how to access external expertise, safety and commercial considerations, and how to carry out the techno-economic assessments. Emphasizing the commercial design and understanding the critical parameters as early as possible are key to reducing cost and time. The second part covers the basics of piloting and scaling up from an engineering perspective, process and equipment considerations, and pilot plant location and operations. Safety is paramount, and implementing hazard identification tools to assess risk throughout the process will also be described.

Chapter 12, "Raising Investment/Financing," addresses various aspects of early-stage company financing. It explores potential sources of funding typically available at different stages of a company's life cycle and what is exchanged during a financing event. Main investment sources, how they are structured, and the pros and cons of each will be discussed. This is followed by a discussion on investment drivers specific to sustainable chemistry technologies as well as key investment impediments unique to start-ups in the sector.

Chapter 13, "Operationalizing a Start-Up Company," provides an overview of key considerations for start-ups with a focus on systems and oversight that every successful new company must put into place. Practical information will be given on how to set up, leverage, and manage effective advisory and director boards. This is followed by a discussion on how to set up and run key operational systems including human resources and health and safety. Finally, setting up effective financial systems will be described with further information on making financial projections.

"Success Stories" is the last part of this book and encompasses Chapter 14, "Making an Impact: Sustainable Success Stories." It contains three vignettes written by entrepreneurs who have lived through some of the commercialization challenges and want to share their experiences. Each contribution summarizes their journey, focusing on the highs and lows and, most importantly, the key lessons that were learned along the way that will likely benefit you. They are intended to serve as testaments to what entrepreneurs in the sustainable chemistry space can achieve when bringing technological advancements to market that deliver economic, environmental, and social benefits.

1.6 Using This Book

There is certainly no requirement that each chapter in this book be read in the order presented. Although the topics are sequenced to build on each other, you are encouraged to skip to sections of most interest at any particular time. Each of the chapters in this book is designed to be used as a stand-alone information source if so desired. In fact, you will notice many individual topics are covered in multiple chapters, although from different perspectives. This is not to emphasize any particular topic or to cause undue duplication, but rather to show that key topics can take on different forms when presented from alternative points of view. For example, an investment professional often perceives key elements of sustainable chemistry commercialization very differently from an intellectual property specialist or scale-up engineer. Furthermore, understanding all these various perspectives allows you to better work with these individuals as you bring your technology through the many stages of development and eventually to market.

You may also notice the differing writing styles of the contributors. As editors, we did our best to keep the tone relatively consistent throughout the book, while giving each author the freedom to express their ideas as they see fit. The keen observer may also note that their individual styles often reflect their specialization – sometimes almost stereotypically so! Engineers, marketing specialists, intellectual property professionals, regulatory experts, academics, and generalists often have writing styles consistent with their professions.

The goal of this book is not to be an exhaustive study of any particular topic, but to provide a broad overview of the challenges of commercializing a novel sustainable technology and practical tools to overcome them. References are included in each chapter for additional details on any particular topic of interest.

Bringing a new sustainable technology to market is a hugely demanding activity and is rife with risks and challenges. However, it is also a noble and exciting pursuit, because beyond the financial gains that can be had (and that is certainly a good motivation), the world needs innovative technologies now more than ever. With hard work, success is possible, and we wish you the best on your journey.

Acknowledgments

We are very grateful to all of the authors who provided their expertise in each chapter. Despite having many demands on their time, they each brought so much knowledge and passion to their respective topics. This book would not exist without their incredible efforts.

References

1 Swift, T., Moore, M., Saifi, Z. et al. (2018). *Elements of the Business of Chemistry*. American Chemistry Council.
2 Oxford Economics. (2019). *Catalyzing Growth and Addressing Our World's Sustainability Challenges*. ICCA.

3 Wadyalkar, S. (2018). 3 megatrends in the chemical industry. MarketResearch.com. https://blog.marketresearch.com/3-megatrends-in-the-chemical-industry (accessed 29 October 2020).

4 Haunschild, R., Bornmann, L., and Marx, W. (2016). Climate change research in view of bibliometrics. *PLoS One* 11 (7) https://doi.org/10.1371/journal.pone.0160393.

5 Sauermann, H. (2018). Fire in the belly? Employee motives and innovative performance in start-ups versus established firms. *Strategic Entrepreneurship Journal* 12 (4): 423–454.

6 Viki, T. (2018). Why large companies continue to struggle with innovation. Forbes Media. https://www.forbes.com/sites/tendayiviki/2018/11/04/why-large-companies-continue-to-struggle-with-innovation/ (accessed 29 October 2020).

7 Schaufeld, J. (2015). *Commercializing Innovation: Turning Technology Breakthroughs into Products*. Apress.

8 Glavič, P. and Kovačič Lukman, R. (2007). Review of sustainability terms and their definitions. *Journal of Cleaner Production* (18): 15, 1875–1885.

9 Anastas, P.T. and Warner, J.C. (1998). 12 principles of green chemistry. In: *Green Chemistry: Theory and Practice*, 30. New York: Oxford University Press.

10 Zhao, W. (2018). Make the chemical industry clean with green chemistry: an interview with Buxing Han. *National Science Review* 5 (6): 953–956.

11 Hauschild, M.Z., Rosenbaum, R.K., and Olsen, S.I. (2018). *Life Cycle Assessment Theory and Practice*. Springer International Publishing.

12 Jolliet, O., Saade-Sbeih, M., Shaked, S. et al. (2015). *Environmental Life Cycle Assessment*. CRC Press.

13 Bisinella, V., Albizzati, P.F., Astrup, T.F., and Damgaard, A. (eds.) (2018). *Life Cycle Assessment of Grocery Carrier Bags*. The Danish Environmental Protection Agency.

14 Clapp, C. (2018). Investing in a green future. *Nature Climate Change* 8: 96–97.

Part I

Laying the Foundation

2

Marketing and Landscape Analysis
Tess Fennelly

GreenSustains, Inc., Minneapolis, MN, USA

2.1 Introduction: Think Marketing

Successfully marketing and commercializing a technology in sustainable chemistry requires the same market-driven discipline needed to commercialize any other new technology. It is critical to understand who the competition is, what customers need, and how you can successfully differentiate your product, service, or process to those customers.

To begin, we need to understand the concept of "value," which is a measure of what your technology is worth. This worth could be defined by monetary terms to the paying customer or some other factor beyond actual cash value. Although it would be ideal to use a formula to precisely calculate this measure, no such formula exists. That said, several factors play a key role in influencing a technology's value, including customer needs, distinctive competencies, and competitive offerings. The sweet spot that defines a customer's definition of value is the intersection of these factors (see Figure 2.1).

Once the value to prospective customers is understood, the work to define and segment markets for the proposed technology can begin. Finding and prioritizing market focus is paramount. The following sections cover understanding and mapping customer needs and then segmenting and selecting market segments. This is the definition of being market driven. This requires making important decisions that are informed by customer information, competitive intelligence, and clarity in the early-stage company's competencies and value proposition.

How to Commercialize Chemical Technologies for a Sustainable Future, First Edition.
Edited by Timothy J. Clark and Andrew S. Pasternak.
© 2021 John Wiley & Sons Ltd. Published 2021 by John Wiley & Sons Ltd.

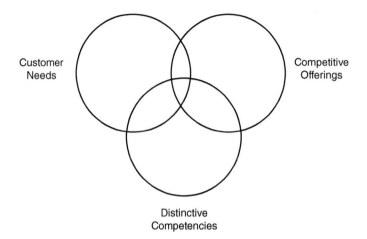

Customer Needs

Competitive Offerings

Distinctive Competencies

Customer needs: Also called pain points, these are what the customer is looking to fill or improve.
Competitive offerings: These are the products/services in the marketplace competing with you and also trying to satisfy the customer needs.
Distinctive competencies: These are unique offerings that you or your competition have that meets the critical market needs better than any of the other available offerings.

Figure 2.1   Factors defining value.

Innovators often confuse "sales" with "marketing." While certainly related, they are not synonymous and have very different mentalities associated with them, as shown in the following table:

Sales approach	Marketing approach
• Focused on the needs of the seller and "selling out" projected or existing inventory • Driven by the seller's need to convert a product or process into cash • Profits driven through sales volume	• Focused on the needs of the buyer • Consumed with the goal to satisfy customer needs • Profits driven through customer satisfaction

As an innovator, you should aim to have your company on a "think marketing" track.

Success (which certainly includes sales) comes as a result of determining the *needs and wants of target segments (a group of customers with a common set of needs)* and then adapting to deliver the desired satisfaction more effectively and efficiently than the competition. This is a marketing approach.

Implementing a marketing approach includes many elements and activities. This chapter will present a conceptual framework (known as an "application framework") to achieve this and offer practical advice for several of the most critical steps. A key "strategic framework" element on selecting market segments is also included. Usable templates are provided so you can apply this process to your early-stage company or technology.

2.2 Creating a Marketing Plan: The Application Framework

This section introduces the application framework for creating a thorough marketing plan. It is the "database" that includes all the information related to customer needs and competitive offerings. The following are the elements of the framework:

Application framework

- Customer needs definition
- Market segmentation
- Market segment evaluation
- Key customer analysis
- Driving forces and life cycle analysis
- Market structure definition
- Industry analysis
- Competitor analysis
- Competitive "product" analysis
- Distinctive competency analysis
- Market research

Once complete, the application framework is used for the strategic framework and future decision-making. The first strategic framework element on selecting market segments and strategic customers is addressed here.

Creating the application framework may seem like a daunting task at first, but it can be readily broken down into a series of manageable steps, as shown in Figure 2.2.

The following sections will examine several of the key elements of the application framework. Sample questionnaires and templates are also included to give innovators a head start in collecting the required information.

2.3 Customer Needs and Mapping

Needs, needs, needs: it may sound redundant, but it is probably the most crucial thing to understand in developing a successful business. You may have a great technology, a strong intellectual property position, and a talented team, but without really understanding the hot buttons of your future customers, you will be doing a "product push," a slow and often unsuccessful approach. Understanding and delivering on identified customer needs will create a "market pull" and the best path to victorious future business.

That being said, finding out these needs isn't always straightforward. There are a number of obstacles and challenges when engaging with a potential customer. For example, you may do any of the following:

- Speak a different language

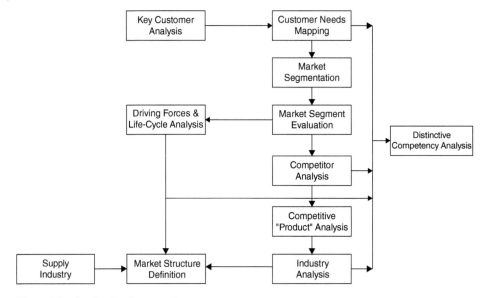

Figure 2.2   Application framework steps.

- Have different knowledge levels
- Impose your own values on the customer
- Have never purchased like a customer
- Not know the customer's business as well as you think
- Not know the customer's need for, and definition of, sustainability
- Think about constraints, not opportunities to satisfy the customer's needs
- Become over-enamored with, and "blinded" by, your technology
- Fail to recognize nontechnical and complex product or process attributes
- Assume your product is more valuable than it actually is

Customer needs are internal conditions that motivate their purchases or use of specific products and services. *What* the customer wants must always be balanced with *why* they want it. Here is the key message:

> We must put ourselves in the customer's shoes to effectively understand their needs.

2.4   Customer Analysis: How to Gather Customer Needs Data

This section will focus on the Key Customer Analysis box in Figure 2.2.

You have an idea of who your customers are, but this may not be true. So, go ahead – get out and talk to your potential customers. The answer is not on your computer, in your lab, or in the conference room. Not until you talk to people will you know whether they are truly a future customer.

This type of research is often called "voice of the customer" (VoC). Many established companies have contracts with external firms to do this work for them. These marketing consultants specialize in running focus groups, mass surveys, primary interviewing, social media mining, etc. While they connect to groups to which the innovator rarely has access, they are necessarily expensive and often beyond the reach of the lean-operating entrepreneur.

Even if you can afford to outsource this work, *don't*. You are commercializing a new innovative sustainable technology in the chemical sector. Do the legwork yourself for now.

Each customer interview or conversation is free education and will contribute to your needs database. It is also the beginning of relationship building in the industry. Talking to potential customers yourself will build your skills and provide insight into industry developments, potential new innovations, technical and/or business direction (often resulting in a change in your roadmap), timelines, and sources of revenue. Although you go into these discussions with the goal of finding specific information, they often take unexpected turns and provide surprising and helpful insights that were simply not anticipated.

You may also be surprised by the willingness of many individuals to openly talk about their business needs with you. You will undoubtedly have some rejections when searching for appropriate contacts, but you will come across many people happy to spend time with you, especially if you openly share your market learnings with them. You must remember that it is often in their interest to learn about the latest emerging technologies. You may actually be helping them fulfill their job requirements as well!

To design, develop, and deliver a product with superior customer value, you must have tools and methods to effectively gather and analyze the associated data. The first tool is an interview form that is conversational and captures the different types of customer needs, shown in here with some examples:

Strategic	Operational	Functional
• Need for product innovation	• Regional/global supply	• Renewable content
• Improved sustainability profile	• Shipping/packaging	• Performance properties (physical, aesthetic, thermal, aging, etc.)
• Global growth	• Central decision-making	• Processability (ease, speed, throughput)
• Importance of service and relationships	• Total cost reductions	• Recyclability
• Importance of price	• Regulatory knowledge	• Environment, health, and safety product profile
• Risk profile	• Environmental services	• Problem correction for end-use
	• Process quality control	
	• Technical and customer service	

Time is limited during an interview, so your questions should quickly get to the core issue of "What does the customer want?" All product attributes and services the customer is seeking should be captured. If the customer does not mention a known attribute that could be critical, it should be raised as a question.

Although most interviews are likely to occur by phone/videoconference, try to have some in-person, "face-to-face" visits too. They are time-consuming and may incur travel costs, but the depth of learning (and extent of relationship building) is far greater.

2.4.1   Finding the Right Contacts

One of the first challenges is finding the most appropriate and helpful contacts with whom to set up conversations. Be creative and persistent in looking for contacts of value. Start with the following ideas and expand if needed:

- Mine your own contact list.
- Ask your network for introductions.
- Search for contacts on LinkedIn and Google.
- Mine the contacts from trade and industry journals and websites.
- Mine conference or trade show attendee lists/panelists.

These are other places to mine for appropriate contacts:

- Customer technical visits
- Sales calls
- Service calls
- Customer complaints
- Industry experts
- Trade journal editors
- Suppliers
- Annual reports
- Websites

2.4.2   The Interview Form

Having a prepared interview form is key to extracting value from a discussion with a potential customer. Obviously, depending on your technology, the questions will be customized to your business, but the sample provided here gives an idea of how to gather data. It can be modified as you see fit.

Sample VoC interview form

Company

Contact

Title

Phone

Location

Location Type

Application Focus

Product Used

Date

Other Comments

There are several important considerations when conducting your interviews:

- Don't start with the company's purchasing department. You have a new technology and are trying to learn if and how it fits with the customer. Their technical and market development teams are generally the first to review new technologies, so you should focus on gaining access to this group. Purchasing will get involved down the road or can be the target of subsequent calls (additional interviews).
- Use your networks to find the first people to interview. During your interview, ask whether the interviewee would recommend others with whom you could speak. Keep branching out.
- Approach it as a conversation, not a formal survey. People generally don't like to participate in surveys but are happy to have informal conversations. Besides, you will learn things you hadn't anticipated by simply talking, rather than surveying. This will more than likely cause you to modify your interview form, which is a good thing. You are already learning what the customers need!
- In your opening, offer to share what you have learned in your discussions to date. This can be an attractive reason for someone to speak with you.
- Purchase volumes, price, and suppliers are often more delicate subjects. Don't start with that. Build the rapport first.
  - If someone is unwilling to share volume data when you ask, try "I understand you don't want to tell me the specific volume, but I'm just trying to gauge magnitude. Do you purchase less than $1 million per year but more than $100,000?," etc.
  - The same goes for pricing. If you have heard about some pricing information in the industry, but they aren't willing to share this information, you can ask if they are paying more than what you heard: above x or under y.
- Document your discussions *immediately*. It's amazing how fast details fade no matter how astute your memory.

The following is an example of a script to use for your interviews. Again, it is a living document to be revised as you gain more information so that you can refine your questions.

> "We are Company Z, which has a new sustainable chemical technology that may have potential as a new X in your market area. We have finished Y testing and have excellent broad-based results. As we move to further testing/product development/commercialization, we are trying to gain an understanding of which application areas may have needs to which we can tailor properties and performance. May I ask you a few questions? I'm happy to share with you what I've learned in speaking with others in the industry."

1. What X do you currently use? In what applications?

   A:

2. How did you choose the materials you use now? Who makes the decision to buy or change?

   A:

3. What are the most important requirements for your X?
   a. Technical
      i. Physical form
      ii. Renewable content
      iii. Efficacy
      iv. Durability
      v. Color
      vi. Odor
      vii. Other
   b. Business
      i. Supplier technical support
      ii. Development needs
      iii. Product reproducibility
      iv. Importance of price
      v. Supplier relationship/reputation
      vi. Formulation/testing
      vii. Delivery: logistics
4. How important is sustainability in your decision-making? What elements of sustainability are most important to you?

   _____

   A:
   _____

5. Are there any areas where you would like some improvements, need additional performance? For which of these, if any, would you be willing to pay more?

   _____

   A:
   _____

6. Pricing is always important, but how critical is the cost of X compared to other elements of the formulation or as a percent of the final product?

   _____

   A:
   _____

7. Who do you consider the best suppliers today? Why? (*If you can get product names and what they pay for it, great.*)

   _____

   A:
   _____

8. What would it take to consider a new X? How much of current X do you use?
   a. Time
   b. Cost
   c. Proven performance (define)
   d. Other?

   _____

   A:
   _____

9. What would it take to consider a new X supplier?

    a. Process for approval
    b. Reputation
    c. Portfolio
    d. End-use knowledge
    e. Other?

    A:

10. Is there any significant trend or issue that you foresee affecting your use of X today or in the near future?

    A:

2.5 Customer Needs Mapping

This section will focus on the Customer Needs Mapping box in Figure 2.2.

Ideally at this stage you have accumulated some useful data that can include any of the following:

- Customer needs, wants, and buying behaviors
- Customer goals, objectives, products, etc.
- Changes in needs, forecasts of changes
- Market trends and changes

Now it's time to use this data and look for patterns with customer needs mapping. This will lay the groundwork for the next critical step, market segmentation. To begin, divide your customer needs into strategic, operational, and functional needs as previously defined.

Take the time to really think through what you heard. What were the patterns and common needs? Some needs may have been high priority for some customers and lower for others, but there are often commonalities. For example, maybe a group of customers had the functional need of compostability as high and others had it as either a low or "nice to have." This is where you should be able to start to differentiate needs for groups of customers.

The needs flow chart you develop (see the sample grid shown in Figure 2.3) will be used for every customer, although again the importance of individual customer needs will likely vary.

When using the grid shown in Figure 2.3 for each customer, rank each need on the 1 to 5 scale:

5 – Critically important
3 – Somewhat important
1 – Not important

The next step is to use your grid to create a "map" for each customer interviewed. To do this, map the needs based on your 1 to 5 ranking, drawing lines from one point to the

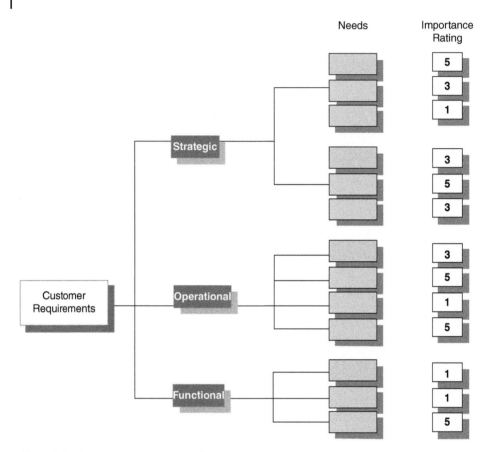

Figure 2.3 A sample customer needs flow chart.

next, as shown in Figure 2.4. Once you have finished, lay out the different maps and start to cluster those that have similar needs.

By understanding customer product and service needs, strategic needs, objectives, and buying procedures, you can do the following:

- Effectively segment the market as it pertains to your technology offering
- Develop products and services that provide value
- Develop targeted positioning and market communications
- Develop effective distribution systems

2.6 Market Segmentation

This section will focus on the Market Segmentation box in Figure 2.2.

What exactly are "target segments"? Target segments are the roadmap for your business. Segmentation is a critical tool as it allows you to address several key questions.

Customer ABC

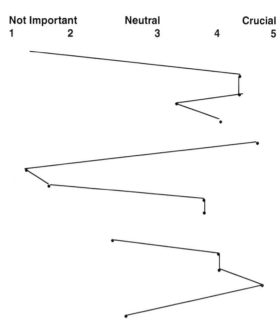

Figure 2.4   A sample customer map for customer ABC.

- You have many potential customers, each with a unique "map" as discussed in the previous section. Do you approach each one with an individual marketing package, or can you create one for multiple customers?
- How do you decide which customers to serve and how to market to them along with which customers not to serve?
- How successful are you at predicting and understanding customer behavior?

Market segmentation is both an analytical and a creative process. It involves the collection and analysis of data regarding the customer and marketplace, but also an imaginative interpretation of that data, as shown in Figure 2.5.

There are many traditional approaches to segmentation as there are many ways to split and categorize the data you obtain after answering the questions in Figure 2.5. For example, you can break out the information based on the following items, each of which can be considered a segment:

- End-use applications (automotive, personal care, building and construction, etc.)
- Company revenue (greater than $2 billion, $500 million to $2 billion, under $50 million, etc.)
- Product (bearings, lubricants, fuel additives, etc.)
- Geography (US, Japan, EU, India)
- Technology (for the example of water resistant clothing: monolithic or microporous film, water repellent coating, inherently water repellent materials)
- Channel position (compounder, distributor, fabricator, machinery manufacturer, etc.)
- Order size (specialty, less-than-truckload, tank car, etc.)

Question	Issues	Activities affected
What	What products, services, and pricing to offer?	Product development, marketing, sales, manufacturing
Who/How	Who to aim at? How to position?	Marketing, sales, planning team
Where/Which	Where in the supply chain should this be positioned? Which marketing channels should be used?	Marketing, sales
Why	Why are customers motivated to buy certain products and utilize certain suppliers?	All

Figure 2.5 Market segmentation questions.

One of these traditional definitions might work for your business, but may fall short in some cases. Referring again to Figure 2.5, a traditional segmentation:

- Does not define the "what" except for the product-based segmentation, including what specific products to offer the customer segment, the price points and approach, need for services and support, etc.
- Does not define the "why" at all. Why are the segment customers motivated to buy your products and utilize you or others as suppliers?
- Covers "where" and "how" only moderately well. Where is the best place for you to be positioned in the supply chain? How will you offer your products/technology? Will it be in a service support and augmented with other products?
- Does not focus on the multidimensional complexity of the buying decision based on functional, operational, and strategic needs that you uncovered during your needs mapping to form customer groupings with similar sets of needs.

A really good segmentation is customized to your business and customer groups based on the needs and needs mapping you have developed.

You are now well on your way to tailored segmentation. You have your graphed customer needs maps and have started to cluster them into groups with common needs (similar map configurations). This is the beginning of your customized segment offering: a needs-based segmentation. It will focus you on sets of customers with similar patterns of needs who should respond to the same set of marketing programs for your product, technology, or service. This will allow you to address the "hot buttons" of these segments (groups of customers) leading to a higher probability of success as you enter the market.

Not all of these segments will be a fit for your company and technology offering. You need to understand whether or not these market segments are attractive: size, growth, profitability, etc. Just because a market segment exists, it doesn't necessarily mean you should spend time and resources pursuing it. You'll also need to understand how you line up with the competition in meeting the customer needs in a given segment. Questions include: Is there an unmet need that you are well positioned to fill? Do you exceed the competition in some areas?

2.7 Market Segment Evaluation

This section will focus on the Market Segment Evaluation box in Figure 2.2.

For this step, look at the customers within each of the segments you have created and see what similarities exist. This will be your initial evaluation of the market segment profile, and the data will be used to prioritize and select your target segments. Areas of commonality may include the following:

- End use
- Geographic coverage
- Size
- Growth
- Customers in the segment
- Supply/value chain
- Trends, dynamics, and driving forces
- Competitive landscape

2.8 Competitive Landscape and Competencies

This section will focus on the Competitor Analysis, Competitive "Product" Analysis, and Distinctive Competency Analysis boxes in Figure 2.2.

During the customer needs analysis, you used Figure 2.3 to profile customers and gave each need a performance rating based on the individual customer's feedback. Then you mapped each customer as shown in Figure 2.4 and clustered them based on common needs to form segments. Now, take the same format as Figure 2.3, but *rank the needs of the segment instead of the individual customer* as done previously (see Figure 2.6). This effectively summarizes the needs of all of the customers in that segment and becomes the segment success factors.

Now, take your top competitors and put them on the chart (see Figure 2.7). Rank how well you believe the competition (competitors A, B, C) does at meeting the segment success factors. You will need to profile the competitors' strengths, weaknesses, and product offerings. Competitor information may be difficult to obtain as it probably isn't in their interest to share too much. Nevertheless, some of this information will likely have been gathered during your interviews or is otherwise available by reading competitor product literature, patents, and other related sources of information. Do your best to fill in the information based on what you know. As you learn more in the future, you can refine your rankings.

Finally, rank your company. In Figure 2.8, you have the following:

- The segment needs, which are the success factors
- The importance of these segment needs according to the aggregate of the customer feedback
- The rating of the competition and their ability to meet those segment success factors
- Your company's ability to meet them

The results of this exercise will start to identify your competencies, strengths, and weaknesses.

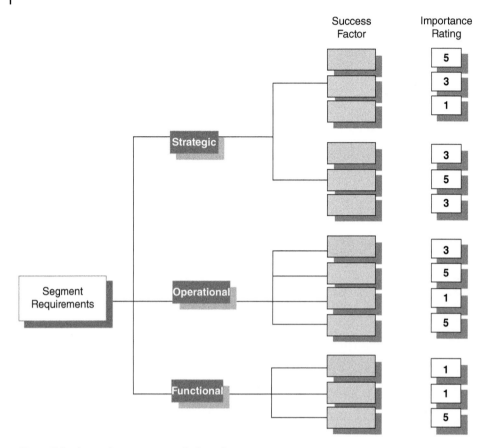

Figure 2.6 A sample segment needs flow chart.

Let's consider each of the three success factor examples provided in Figure 2.8 to demonstrate how you would interpret the results to identify your company's weaknesses, strengths, and distinctive competencies.

- **Strategic**: A top strategic success factor given here is growth from the base. Customers are often given a directive to grow their base business by a certain percentage in a year and to meet certain market share targets. If a new technology will help them do that, then this is a success factor. The segment ranked this a 5 making it a critical success factor. Neither you nor your competitors are meeting this success factor. To meet the needs of customers in this segment, you need to offer a technology that will allow them to grow from the base.
- **Operational**: An operational success factor is to reduce costs. The segment ranked this a 5, also making it a critical success factor. Both you and competitor A are meeting this need. This is excellent in terms of position, but it is not a distinctive competency for either of you since you are both meeting the market need; it is a core strength.

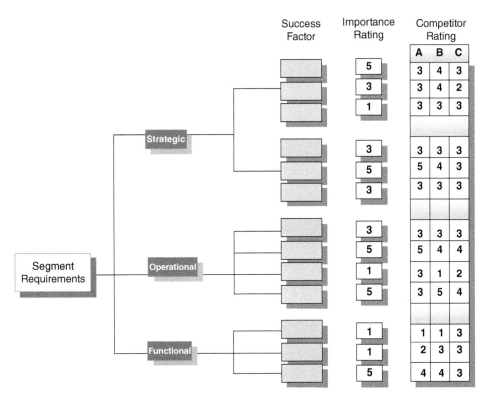

Figure 2.7 Segment importance ranking and competitor rating.

- **Functional**: A critical functional success factor (ranked 5) is renewable content in the product. You are the only one of the competitors to be meeting this need (circled in Figure 2.8). This is a distinctive competency for your company. It is one that you can leverage to gain market penetration and market share.

Once segmentation is completed and your strengths, weaknesses, and competencies are identified, you can prioritize the segments. These can be prioritized according to the following:

- Segment attractiveness
- Customer needs alignment with your competencies
- Competitor strength

2.9 Conclusion and Next Steps

Completing this application framework will identify and define your segments based on clusters of customers with common needs and inform which segments your company should target first. It will also pinpoint which core strengths and competencies you can

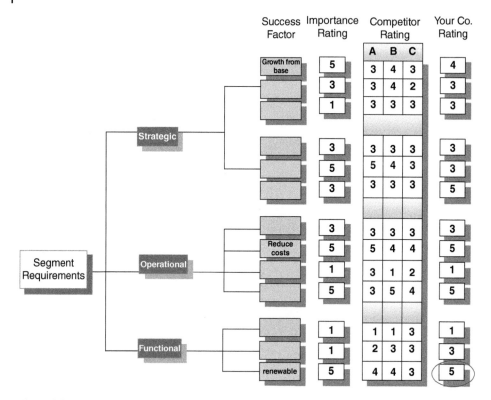

Success Factor	Importance Rating	Competitor Rating			Your Co. Rating
		A	B	C	
Growth from base	5	3	4	3	4
	3	3	4	2	3
	1	3	3	3	3
Strategic					
	3	3	3	3	3
	5	5	4	3	3
	3	3	3	3	5
	3	3	3	3	3
Reduce costs	5	5	4	4	5
	1	3	1	2	1
	5	3	5	4	5
	1	1	1	3	1
	1	2	3	3	3
renewable	5	4	4	3	5

Figure 2.8 Ranking your company.

leverage and which critical market success factors you need to improve upon for greater success. Selecting target segments (based on representing value to your business and technology) will provide focus and guide strategic investment decisions in a wide number of areas, including the following:

- Technology development
- Manufacturing
- Marketing
- Sales
- Distribution

The application framework also offers several other benefits. For example, it allows you to minimize the number of discrete marketing packages to be prepared and to tailor them so that they hit your potential customers' "hot buttons." It focuses efforts on groups of customers whose needs are compatible with your technology offering, which in turn ensures higher hit rates at lower costs. Finally, it helps anticipate and prepare for competitive reactions.

In conclusion, completing these exercises will result in a roadmap targeting the intersection of customer needs, distinctive competencies, and competitive offerings (Figure 2.1). This point of intersection should now be more defined and the value of your technology better understood.

Moving forward, the framework can then be used to create a strategic framework, as shown here, which is vital for future decision-making:

Strategic framework

- Selecting market segments and strategic customers
- Positioning in the marketplace
- Pricing strategy
- Relationship marketing
- Global marketing and analysis
- Distribution strategy
- Marketing strategy
- Marketing organization
- Competitive advantage
- Organizing and implementing the marketing effort

These are more advanced concepts that can be addressed as you develop your business and strategy.

3

Determining the True Value of a Sustainable Chemical Technology

Lauren Heine[1] and Margaret H. Whittaker[2]

[1] *ChemFORWARD, Spokane, WA, USA,*
[2] *ToxServices LLC, Washington, DC, USA,*

3.1 Introduction

Capitalizing on a sustainable chemical technology requires that you embrace the concept of value. In business, the value of an article, technology, or service is often thought of as its market value, which when expressed in terms of money, is called price. "Market value" can be distinguished from "value in use," or the utility of an item, service, or technology for satisfying consumer wants or needs. In a classic observation by Bridgewater and Sherwood [1], value in use is subjective. For example, a loaf of bread's subjective value is much greater to a starving person than to a person who has just eaten a complete meal, noting that the loaf of bread can be bought by either individual at the same price.

Ideally, a technology's market value *and* value in use will increase over time, enjoy widespread adoption, and confer net benefits to the entrepreneur, society, and the environment at large. Throughout history, groundbreaking innovations have given us modern conveniences such as electronics, refrigerants, pharmaceuticals, and low-cost access to an immense array of industrial, agricultural, and consumer goods. These innovations, more often than not, have been accompanied by unintended consequences such as toxic chemicals released to air, water, and the food supply as well as the generation of toxic wastes that are difficult to manage or dispose. These call into question the value of the actual innovation to humankind and the planet.

Most inventors dream of being disruptors and yearn to change the way things are made on a global scale. This is exemplified by the exponential and sustained increase in global production of polypropylene using Ziegler-Natta technology, for which Ziegler and Natta were awarded the Nobel Prize in Chemistry in 1963 [2]. Unfortunately, market dominance does not equate to sustainability. As shown in Figure 3.1, the Great Pacific Garbage Patch (now floating in the ocean in two distinct patches twice the area of Texas) comprises an estimated 1.8 trillion pieces of plastic (largely polyolefins including polypropylene), choking marine life through entanglement or ingestion and impacting our food supply through the biomagnification of pollutants [3–6]. Clearly, we can and must do better.

How to Commercialize Chemical Technologies for a Sustainable Future, First Edition.
Edited by Timothy J. Clark and Andrew S. Pasternak.
© 2021 John Wiley & Sons Ltd. Published 2021 by John Wiley & Sons Ltd.

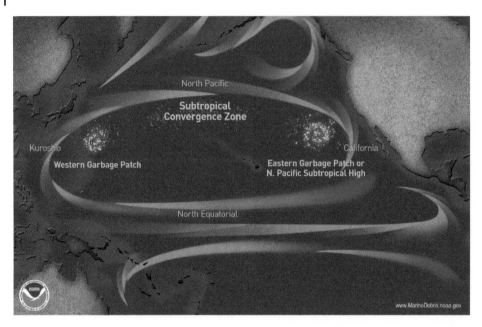

Figure 3.1 Great Pacific garbage patch. Source: National Oceanic and Atmospheric Administration (NOAA), 2012. [3].

For a chemical technology to be sustainable, it must not only have market value and value in use, but must also confer benefits to human well-being and the environment. Sustainability is a complex, systems-based concept that is not as easy to measure as monetary value. But it is important nonetheless to define sustainability for chemical technologies, to set aspirational goals, and to measure progress toward achieving those goals. This chapter will introduce valuation concepts, tools, and examples as they relate to sustainable innovation. We will illustrate how products that confer benefits to human well-being and the environment can address some key business risks, while benefitting from increased value in use.

3.2 Sustainable Value and the United Nations Sustainable Development Goals

What do we mean by "value" to human well-being and "value" to the environment? That is not a simple question. Many well-meaning individuals along the supply chain exert themselves with the intention to benefit both human well-being and the environment through their product design, development, and/or purchasing practices. However, it is important to confirm whether their efforts will truly realize these benefits compared to other products that may provide the same service. Progress may entail incremental improvements or breakthrough innovations. But progress is discernible only by benchmarking against sustainability principles using standardized tools and metrics.

Figure 3.2 The United Nations Sustainable Development Goals. Source: The United Nations Sustainable Development Goals. © 2020 United Nations.

In this chapter, we put forward the United Nations (UN) Sustainable Development Goals (SDGs) as the "North Star" of consensus-based goals reflecting global sustainability principles that can guide chemical technology innovation (Figure 3.2).

According to former UN Secretary-General Ban Ki-moon, "The seventeen sustainable development goals are our shared vision of humanity and a social contract between the world's leaders and the people" [7]. The SDGs were adopted in 2015 by 193 heads of state and other top leaders as part of the 2030 Agenda for Sustainable Development [8]. While priorities will continue to shift in subsequent years due to political change and unanticipated events like the COVID-19 pandemic, the SDGs still reflect a shared vision of human well-being [9]. The UN website provides supporting facts and figures, targets, and links for each of the SDGs.

You should understand how your chemical technology supports or does not support the 17 SDGs. The extent to which your technology moves beyond the status quo toward achievement of the SDGs marks the extent to which the innovation is more sustainable than the status quo. This, in turn, impacts its value with respect to sustainable development. Not all of the 17 SDGs will necessarily be relevant to an innovation, but the exercise of benchmarking is essential.

Each chemical technology is unique, and there is no single formula for measuring sustainable value. It's up to you as an entrepreneur committed to sustainable innovation to decide how best to design your product or process to advance the SDGs, and which human health or environmental impacts you want to address. Early in the technology development phase, you should explicitly define your sustainability goals. What service is your product/process providing? What currently provides the same service? What are the sustainability impacts of current products and services? You can use the SDGs to help you innovate and improve on incumbent products or processes.

3.2.1 Embracing SDGs at the Business Level: United Nations Global Compact Participation

In 2000, the UN Secretary General created the UN Global Compact to align business operations and strategies in the areas of sustainable development. To date, more than 12 000 businesses and 3000 affiliated organizations in 160 countries have participated and learned how to assess, define, implement, measure, and communicate a sustainability strategy. Companies joining the Global Compact have access to sustainability case studies, tools and guidance, and training [10].

The Global Compact administers two SDG recognition programs that award excellence at the individual level: one program acknowledges contributions from established business leaders, and one program trains and fosters development of up-and-coming sustainability leaders. The latter, the Young SDG Innovators Program, is a 10-month accelerator designed for rising future business leaders under-35 who work to develop new technologies, initiatives, and business models within a Global Compact participating organization. The 2020 program comprised 480 participants from 130 companies delivering more than 80 SDG business solutions.

3.3 Life-Cycle Thinking and Life-Cycle Assessment

Sustainability is measured at multiple scales. To ensure that what may appear to be an innovation at one scale will impart appropriate sustainability benefits at other scales, it is essential to consider the full life cycle of a chemical technology. A product life cycle typically starts with extraction of resources, then production and manufacturing, transportation, use, and end-of-life management, as illustrated in Figure 3.3.

Figure 3.3 Inputs into product life cycle. Source: Based on Wordclouds.com. 2020., WordCloud Generator Tool. [11].

We distinguish here between "life cycle thinking," an approach that considers aspects associated with the full life cycle of a chemical technology, and a tool called "life cycle assessment" (LCA). Life cycle thinking and LCA may be closely related, especially when the parameters being evaluated are those included in the LCA methodology. However, it is possible to apply life-cycle thinking to evaluate impacts that are not typically part of the LCA methodology. For example, LCA typically measures impacts associated with material inputs and outputs and energy consumption. One might also consider the social impact of exposure to skin-sensitizing substances or the likelihood of generating litter across the full product life cycle.

The SDGs are high-level, long-term goals. To achieve progress, these can be broken down into aspects and impacts that will allow you to benchmark the sustainability benefits and trade-offs associated with your innovation. Aspects and impacts are terms borrowed from ISO 14001 Environmental Management Systems. An "aspect" is an element of your company's activities, products, or services that can interact with the environment, whereas an "impact" is the resulting change to the environment. The impact can be adverse or beneficial, and wholly or partially resulting from the aspects [12]. Another term we will use here is product "attributes." Attributes refer to inherent features of products that are linked to aspects and impacts. An example is recyclability, which can be an attribute of a material or product that is linked to the aspect of waste generation and intended to reduce negative impacts from wastes. Examples of impacts that are well accepted in LCAs include ozone depletion, depletion of fossil fuels, emission of small particulates, emission of carcinogens, and eutrophication.

3.4 Attributes and Impacts: Check Your Assumptions

A report by the Oregon Department of Environmental Quality (DEQ) with Franklin Associates provides great examples of how relying on product design attributes alone can provide an incomplete picture of a product's environmental impact [13]. While attributes can be useful design criteria, it is critical to recognize that attributes do not necessarily describe an actual positive environmental outcome.

For example, the report evaluated how well four popular packaging attributes (recycled content, biobased, recyclable, and/or compostable) correlated with net environmental benefit across the full life cycle of packaging, from resource extraction to manufacture, distribution, use, and disposal (Figure 3.4).

The analysis gleaned some interesting insights into the difference between making decisions based on environmental attributes versus actually measuring environmental impacts. For example, when using packaging made of the same material (e.g. polyethylene terephthalate [PET] plastic), the option with higher recycled content generally had lower overall environmental impacts than the option without recycled content, as perhaps expected. However, that finding did *not* hold when comparing between materials. Recycled glass did not necessarily have better overall environmental performance than virgin PET. Therefore, Oregon DEQ recommended against using these attributes as a rule of thumb when comparing *between* materials.

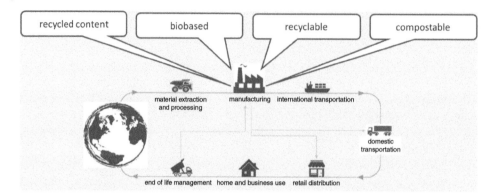

Figure 3.4 Case study of examining the correlation between packaging attributes and net environmental benefit across a full life cycle. Source: image from Oregon DEQ website. Source: Oregon Department of Environmental Quality (DEQ), 2018 [14].

Likewise, they learned that the benefits of biobased content depend on agricultural production methods and associated impacts on local communities. As such, the authors determined that "biobased" as an attribute was unreliable for predicting health and environmental benefits, which instead need to be confirmed on a case-by-case basis. Recyclable materials can be beneficial as long as there is high potential for them to be recovered from waste streams and actually recycled. But even then, it is possible that there are other aspects and impacts that result in undesirable trade-offs. For example, the manufacturing process to produce a recyclable material could involve toxic chemical emissions and high energy consumption.

Compostable materials can also have benefits, but like the recyclable attribute, this depends on the availability of composting infrastructure. It is important that these materials used in products fully compost in dedicated facilities and that they do not contaminate the final compost. If the compost produced is intended to have value to those who use it, then perhaps composting should not be seen as a means of disposing biodegradable wastes, but rather as a process for creating beneficial soil amendments.

Figure 3.5 illustrates the relationship between attributes and environmental impacts. Although each of the four packaging attributes shown in the figure are laudable design goals, designing a packaging material that fulfills one or more of these attributes does not mean that the packaging will have the positive impacts shown at the far left of Figure 3.5.

The previous examples are meant to drive home the importance of taking a life-cycle perspective and for checking assumptions about how product attributes will impact human health and the environment. It is important to note that LCA is an excellent tool for optimizing efficiencies in material and energy use, but it may not always address what is important to society and align with the SDGs. For example, a community may value job creation over energy efficiency, or waste management policies that drive material takeback and product stewardship over ongoing use of landfills.

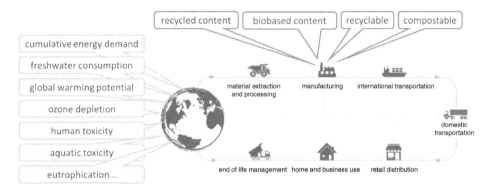

Figure 3.5 Interrelationship between attributes and environmental impacts along a full life cycle. Source: Oregon Department of Environmental Quality (DEQ), 2018. Technical Summaries [15].

3.5 Business Risk and Sustainable Design – Or How to Turn an Externality into a Selling Point

Victor Papanek's 1971 book *Design for the Real World: Human Ecology and Social Change* is credited with advancing the concept that industrial designers not only have a responsibility to themselves but to the world at large [16]. He calls out the fact that nearly all major disfigurements of Earth have been created by humans. Papanek's teachings apply equally to the sustainable chemical technology entrepreneur.

As noted, impacts can be costs or benefits, can be positive or negative, and can affect you (the entrepreneur) or society as a whole. An "externality" is a cost or benefit caused by a producer of a chemical technology that is not financially incurred or received by that producer. Classical economic theory states that, by definition, an externality does not affect the entity that causes the externality. For example, a manufacturer of single-use products or packaging that soon becomes waste may not incur the cost of managing that waste. However, sometimes externalities come back to bite you and become business risks. Being proactive about sustainability externalities can lead to positive selling points.

Business risks are a good place to look for opportunities for sustainable chemical technology innovation. Taking actions that reduce human health and/or environmental impacts *and* business risk is like picking the proverbial "low-hanging fruit," especially when those impacts are linked to "value in use" to a consumer. Addressing sustainability impacts as a business risk is a conservative way to integrate sustainability into product design – to avoid a situation where the market value may not reflect real sustainability benefits.

One classic example of business risk in product design and development is the use of chemicals that have high inherent hazards of concern to consumers, but that are legal to use. Sometimes awareness of a chemical's hazards reaches the scientific community, nongovernmental organizations (NGOs), or the general public before regulators take action as described in Chapter 6. While it is possible to keep using the chemical until it becomes restricted or banned, more forward-looking companies seek to avoid negative public perceptions and keep ahead of future regulatory controls by looking at the hazard attributes of their product's chemicals.

Bisphenol A (BPA) is a good example of such a business risk. BPA was first synthesized in 1891, and in the late 1930s scientists discovered that it acted as an artificial estrogen. In the 1950s, BPA began to be used widely in the manufacturing of plastics and to appear in consumer products throughout the world. Despite its origins as a potential pharmaceutical, the timing of its development allowed it to be "grandfathered in" and to be presumed safe under the Toxic Substances Control Act (1976) [17].

When the general public became increasingly concerned about exposure to BPA from water bottles and other products such as thermal paper receipts, some manufacturers sought to avoid BPA by selecting the next best candidate – one that would work similarly to BPA but that was not BPA. An example of such an alternative is BPS, which is a structural analog of BPA. BPS was less well known and less well studied, and therefore there was little data available showing that it is toxic or endocrine disrupting. Toxicologists use a technique called "read across" to predict the toxicity of a less well-known chemical by reviewing the toxicity of a structural analog with more plentiful data. The best structural analog for BPS identified by the US EPA for use in thermal paper receipts was, in fact, BPA [18]. A toxicologist would never advise the substitution of BPA with BPS based on concerns for toxicity. Substituting a chemical of concern with a chemical that has similar, or even dissimilar but undesirable hazard properties, is called a "regrettable substitution." Resources for designing safer molecules are available that can help chemists follow basic principles to make chemicals less bioavailable and to avoid the use of known toxicophores [19]. Safer alternatives to the use of a chemical of concern do not necessarily need to be another chemical. The use of BPA in water bottles can be eliminated by using different plastics made without toxic or endocrine active substances or even nonplastic alternatives such as aluminum.

A second example of not fully assessing product risk during product design is the development of refrigerants that harm humans or the environment. As summarized in Table 3.1, refrigerants evolved from those directly posing hazards to humans to those that harm humans through the environment, such as through depletion of stratospheric ozone or contribution to global warming. The impetus behind refrigerant changes was instigated by an industry-wide desire to find nontoxic, stable, and nonflammable refrigerants [20–22]. One of the inventors of Freon formulations, Thomas Midgley of the Frigidaire Corporation, famously introduced the scientific community to the Chlorofluorocarbon (CFC) R-12 at an American Chemical Society meeting in 1930 by inhaling a lungful of the gas and blowing out a lit candle [23, 24]. Little did the inventors of CFC refrigerants realize that CFCs leaking from refrigeration and air-conditioning equipment would deplete the ozone layer within a short span of time ($\sim$50 years), leaving the earth exposed to potentially harmful ultraviolet radiation that can cause skin cancer and cataracts [25, 26].

The previous examples illustrate that a regrettable substitution may be a chemical with the same hazard properties as the chemical of concern or one that poses very different hazard properties; instead of harming humans, it may harm the environment, or vice versa. That is why knowing the hazard profile of chemicals used in products and processes is so important.

Table 3.1 Human health and environmental hazards posed by commercial refrigerants over the past century

Refrigerant classes and example(s)	Timeframe of use	Specific hazards
Early refrigerants such as NH_3, CO_2, CH_3Cl, and SO_2	1830s–1920s	Flammability, Acute Toxicity, Asphyxia, Death
Chlorofluorocarbons (CFCs) such as R-12 (aka dichlorodifluoromethane, Freon-12®)	1928 – Phase-out started in mid-1990s in industrialized countries	Ozone depletion, Greenhouse gas
Hydrochlorofluorocarbons (HCFCs) such as HCFC-22 (aka Chlorodifluoromethane, R22)	1980s – Phase-out by 2030	Ozone depletion, Greenhouse gas
Hydrofluorocarbons (HFCs) such as HFC-23 (aka Fluoroform, R-23)	1990s – Gradual reduction (80–85%) by the late 2040s	Greenhouse gas

3.6 Guiding Principles for Sustainable Chemical Technology Innovations: Chemistry, Carbon, and Circularity

Entrepreneurs commercializing sustainable chemistry technologies are no doubt familiar with the Principles of Green Chemistry and the Principles of Green Engineering. These principles are effective in guiding the design and development of chemical products and processes that reduce or eliminate the use or generation of hazardous substances [27].

The Organisation for Economic Co-operation and Development (OECD) defines sustainable chemistry as "a scientific concept that seeks to improve the efficiency with which natural resources are used to meet human needs for chemical products and services" [28]. Sustainable chemistry is a newer, broader, and less well-defined concept that provides additional scope and flexibility but only if it incorporates the principles of green chemistry. Sustainable chemistry is not sustainable if natural resource efficiency is gained at the cost of increasing the use and generation of hazardous substances. Therefore, we refer to green *and* sustainable chemistry. We refer you to a resource containing useful metrics and illustrative examples of green and sustainable chemical innovations ranging from molecular design to process design [29].

3.6.1 Sustainable Materials Management

The OECD is a global organization that works with governments, policy makers, and citizens to shape policies that foster prosperity, equality, opportunity, and well-being for all. OECD leads collaborations designed to establish evidence-based international standards and to find solutions to a range of social, economic, and environmental challenges. OECD serves as a knowledge hub for data and analysis, exchange of experiences, best-practice sharing, and advice on public policies and international standard-setting. This includes maintaining an online green growth "hub" designed to foster "economic growth and development while ensuring that natural assets continue to provide the resources and environmental services on which our well-being relies" [30].

OECD leads an initiative called Sustainable Materials Management (SMM). "SMM is an approach to integrating policies and practices that promote sustainable materials use by reducing negative human health and environmental impacts and preserving natural capital throughout the life cycle of materials, taking into account economic efficiency and social equity" [31].

SMM integrates the principles of green and sustainable chemistry and engineering along with principles intended to lead to a "circular economy." Aspiring to a circular economy is sometimes referred to as "circularity."

3.6.1.1 What Is a Circular Economy?

Looking beyond the current take-make-waste extractive industrial model, a circular economy aims to redefine growth, focusing on positive society-wide benefits. It entails gradually decoupling economic activity from the consumption of finite resources and designing waste out of the system. It is underpinned by a transition to renewable energy sources and is based on these three principles:

- Design out waste and pollution.
- Keep products and materials in use.
- Regenerate natural systems [32].

Figure 3.6 illustrates a systems view of material flows through ecological, industrial, and societal systems. Typically, policies and regulations are designed to drive desired practices and to minimize impacts between just two systems at a time. As shown in the figure, natural resource policies address the flow of materials from ecological systems to industrial systems; product life-cycle policies address the flow of products and services from industrial systems to societal systems; and waste management policies address the flow of materials from societal systems back to ecological systems. When policies are not integrated between all three systems, they create policy misalignments that can drive undesirable impacts. For example, policies designed to increase recycling might result in increased toxic emissions from recycling facilities or increased concentration of toxic substances in recycled materials, if toxic substances were used in the original products. Policy integration and systems thinking is essential for SMM and circularity. You should be aware of materials management policies in the regions where you intend to sell your products and understand how these policies will impact options for circularity throughout your product's life cycle.

3.6.2 Alternatives Assessment

Alternatives assessment (AA) provides a structured approach to assessing distinct attributes, aspects, and impacts associated with products and processes. First, each aspect/impact is assessed individually; then a decision analytic framework is applied to compare the products and processes based on the collated aspects and impacts. AA can serve as a helpful resource by taking a holistic approach to product design and development and systematically comparing your product to other products that provide the same service.

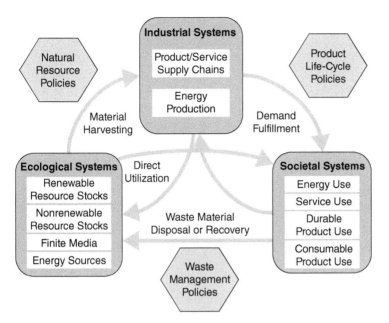

Figure 3.6 Systems view of material flow cycles and policy frameworks. Source: Fiksel, Joseph, A Framework for Sustainable Materials Management, JOM, Aug 2006. © 1969 Springer Nature [33].

An example of a relatively simple alternative to a chemical of concern comes from medical devices that use polyvinyl chloride (PVC) plastic for certain sensitive applications [34]. PVC is hard and brittle at room temperature and typically needs a plasticizer to make it flexible. The most common plasticizers are ortho phthalates such as di-(2-ethylhexyl) phthalate (DEHP). Numerous studies have demonstrated that DEHP may have adverse reproductive and developmental effects, both in animals and in humans. Fortunately, safer alternatives such as di-(2-ethylhexyl) terephthalate (DEHT) can be substituted for DEHP for many applications.

An example of avoiding chemicals of concern by using alternative materials comes from the leather industry. The leather tanning process contributes to global warming, uses highly toxic chemicals such as chromium salts and corrosive chemicals, and pollutes water [34]. An alternative to the use of toxic chemicals in the tanning process is to use "leather" derived from mushrooms. Mylo™ leather is produced from mycelium cells that can be grown in days versus years as is the case with animal hides. According to reviews, it is remarkably "leather-like," supple, durable, and abrasion resistant but without the environmental impacts associated with tanned, hide-based leather [35].

An example of avoiding chemicals of concern through alternative product designs comes from the electronics sector. Whenever flammable materials such as plastics are combined with high energy or heat sources such as laptops, flammability becomes a hazard. Halogenated organic flame retardants such as polybrominated diphenyl ethers (PBDEs) were commonly used to prevent ignition and flame propagation. Industry standards require flame retardancy testing. Appropriate flame retardants are tied to test requirements and compatibility with specific plastics. Safer alternatives are not always available. In some cases, fire safety can be achieved by encapsulating the flammable material or by using barriers to separate it from potential exposure to the heat source. This separation can be an effective way to engineer fire-safe products without adding toxic flame-retardant chemicals [36].

The most comprehensive, step-by-step guidance available for AA is the Interstate Chemicals Clearinghouse Alternatives Assessment Guide v1.1 (IC2 AA Guide) [37]. It describes how to evaluate and compare products based on individual modules or impact categories, along with a section on decision frameworks. Each module is defined by several levels that allow the user to use them in a progressive and iterative way. For example, a Level 1 performance assessment may simply require evidence that a product is available and used in the marketplace. A higher-level performance assessment may entail extensive materials testing against industry standards. Lower-level assessments are less data intensive and may be sufficient for decision-making. The following are the modules/impact categories in the IC2 AA Guide:

- Chemical hazard assessment
- Exposure
- Cost and availability
- Performance
- Life-cycle thinking/life-cycle assessment
- Materials management
- Stakeholder engagement
- Social impact assessment

3.7 Chemical and Material Considerations that Impact Sustainable Value

Based on the current landscape of principles, frameworks, tools, and data resources, we advocate that you use a relatively simple approach to designing and benchmarking the sustainability value of your chemical technology. We recommend focusing primarily on the chemicals and materials used across the product life cycle from the perspective of:

- Chemistry
- Carbon
- Circularity

Chemistry, carbon, and circularity are necessary but not sufficient parameters for sustainability. They can be used in conjunction with benchmarking against the SDGs and with economic and market analyses. "Chemistry, carbon, and circularity" is a memorable catchphrase to make sustainable value attributes, aspects, and impacts more manageable (M. Davies, personal communication).

3.7.1 Chemistry

By "chemistry" we refer to the inherent toxicity or other hazards of chemicals in products and the potential for exposure to them across the product life cycle. It is important for you to (i) know the identity of all chemicals used in your product and/or process and (ii) to understand the hazard profiles of those chemicals, including those potentially formed upon release to the environment (environmental transformation products).

Regulations, particularly those modeled after REACH in the EU, provide clear indicators of hazardous properties that are particularly undesirable. They include carcinogenicity, mutagenicity, reproductive and/or developmental toxicity, endocrine disruption, and toxicity to humans or other organisms in combination with persistence, bioaccumulation potential, or mobility in the environment. Chemicals with any of these hazards are considered substances of very high concern (SVHCs) and should be avoided as you may otherwise be forced to replace a chemical in your product based on public concern and/or regulation. Being forced to remove or replace a key chemical in your product that is already launched in the marketplace is a major risk to your business. Substitutes are not always readily available and retroactively removing a chemical from one's supply chain can be a costly nightmare, not to mention the negative public perception.

A growing number of retailers and industry leaders are responding to the public's desire for safer chemicals in consumer products (as detailed in Chapter 6). They are less interested in arguing that toxic chemicals in products are sufficiently safe, and more interested in avoiding their use altogether. They are setting policies that restrict the use of chemicals of concern, even if they are not (yet) regulated. Lists of chemicals compiled to support these policies are called "restricted substances lists" (RSLs). When applied not only to products but also to manufacturing processes, such lists of chemicals are called "manufacturing restricted substances lists" (M/RSLs). Well-known large companies with public RSLs include Apple, Target, Sephora, Nike, and others [38]. Many leading manufacturers and brands have internal procurement policies that restrict the use of chemicals of concern in their products, above and beyond regulations. You would be unwise to think that meeting minimum regulatory requirements is sufficient for your product, which you hope will be in the marketplace in numerous jurisdictions for a long time.

As you develop your sustainable chemical technology you should think proactively and globally about how to avoid business risks from chemical hazards. Drivers for chemical restrictions include regional, national, and international regulations; RSLs and M/RSLs created by potential customers; and NGO campaigns. It is important to have as much information as feasible about the hazard profile of every chemical used in your product or process, perhaps starting with those chemicals most likely to result in exposure to humans or the environment. You should look globally for chemical hazard classification information, as sometimes different countries emphasize different data sets resulting in different hazard classification results. It can be strategic to adopt the most conservative of the authoritative assessments globally. That way, you will not limit your global market access options. Freely available resources such as ChemSafetyPro's GHS Basics and GHS Classification articles [39] compile much of this critical information. Healthy Building Network's Pharos database [40] is another freely available resource that allows for screening chemicals against authoritative lists of chemicals known to have certain hazards.

Increasingly, comprehensive chemical hazard assessment (CHA) reports designed for nontoxicology experts are available for chemicals being used in products that can readily inform decision-making. Methods for evaluating and classifying hazards include the Globally Harmonized System of Classification and Labeling (GHS) [41], Cradle to Cradle

Table 3.2 Hazard endpoints included in leading CHA methodologies

Globally Harmonized System for Classification and Labeling of Chemicals (GHS)	GreenScreen for Safer Chemicals (GreenScreen®)	Cradle to Cradle Certified™ Material Health Methodology (C2CC) as presented in ChemFORWARD
Acute mammalian toxicity	**All endpoints found in the first column plus the following endpoints:**	**All endpoints found in the first column plus the following endpoints:**
Skin corrosion/irritation		
Serious eye damage/eye irritation		
Respiratory sensitizer	Persistence	Persistence
Skin sensitizer	Environmental transformation products	Exposure to parent chemical or transformation products across the product life cycle
Germ cell mutagenicity	Bioaccumulation potential	
Carcinogenicity	Endocrine activity/disruption	Bioaccumulation potential
Toxic to reproduction (includes developmental toxicity)		Endocrine activity/disruption
Specific target organ toxicity following single exposure		Climatic relevance
		Toxic metals
Specific target organ toxicity following repeated exposure		Organohalogens
Aspiration hazard		Terrestrial toxicity (avian, earthworm)
Hazardous to the aquatic environment, short-term (acute)		Other human (i.e. skin penetration, phototoxicity VOCs)
Hazardous to the aquatic environment, long-term (chronic)		Other environmental (i.e. mobilization of metals, other aquatic impacts)
Hazardous to the ozone layer		
Various physical properties: flammability		
reactivity		

Certified™ Material Health Methodology [42], GreenScreen for Safer Chemicals® [43], and the US EPA Design for the Environment Alternatives Assessment Criteria for Hazard Evaluation [44]. Most methods are based primarily on GHS with the addition of various hazard endpoints, as illustrated in Table 3.2.

Repositories of CHAs can be found via nonprofit platforms such as the ChemFORWARD Globally Harmonized Repository [45] and the Interstate Chemicals Clearinghouse Chemical Hazard Assessment Database [46] or via commercially available databases including ToxServices' ToxFMD Screened Chemistry® Library [47], SciVeraLENS [48], and Verisk 3E [49], among others. Some of these resources also provide information on whether a chemical is found on global regulatory lists, authoritative hazard lists, or industry M/RSLs. Lists are important leading indicators of business risk. Beyond list screening, gathering actionable comprehensive information on chemical hazards for all chemicals used in your technology will further reduce business risk.

Data are rarely available for all hazard endpoints for every chemical. It is important to understand what is known and not known about the hazard properties of a chemical. Some of the CHA methods have scores based on algorithms that take into account data gaps. But one should consider those scores or benchmarks with caution because the significance of a data gap depends on how the chemical is used and how exposure is likely to occur. For example, a chemical may have low toxicity based on oral exposure, but no data when it comes to inhalation exposure. If the chemical is likely to be inhaled during use, then low hazard via the oral exposure route should not necessarily be extrapolated to the inhalation exposure route. Just consider flavors or other chemicals used in vaping.

Where publicly available data are not available, you should request data on the chemical from the supplier. If you are the chemical manufacturer, then every effort should be made to comprehensively evaluate the inherent hazards of the chemical. In addition to obtaining information from global chemical hazard databases and regulatory resources, the use of new approach methods (NAMs) is highly recommended to help predict inherent hazards without the use of animal tests that are costly and increasingly seen as inhumane. Examples of NAMs include read across, structure activity modeling, and *in vitro* testing. While toxicological testing may be required prior to registering a chemical for use in the marketplace, you can do your best to ensure you are selecting the safest possible chemicals early in the development process, using publicly available data and NAMs. This will reduce the business risk of finding unwanted hazard properties during regulatory reviews. Unless you are a toxicologist, such advanced assessment will likely need to be commissioned from a professional toxicology firm.

Minimum actions for you to take to ensure the use of safer chemicals in your product or process include (i) obtaining full material disclosure and (ii) ensuring that you have enough data on all chemicals made or used to prepare a GHS-compliant safety data sheet (SDS). Full material disclosure means knowing all of the intentionally added chemicals and any impurities or residuals above a low concentration threshold (e.g. 100 ppm). If you engage with overseas contract manufacturers, use a strategy of "trust but verify" through affidavits and testing that the chemicals used in their products are those that were specified. Cases where even explicitly unwanted chemicals have infiltrated the supply chain have led to recalls. For example, despite being banned in the US in 1978, lead paint is still used in other countries and has been found in children's toys imported to the US [50]. Obtaining full material disclosure can be time-consuming, especially if your supplier is concerned about disclosing confidential business information. Once you have this disclosure, there are resources to help you generate a GHS-compliant SDS [51]. If you are unable to create your own, there are numerous third-party providers that can generate one for you for a fee, typically starting at ~$200 [52].

More aspirational actions build on the minimum actions described earlier and may cost several thousand dollars. Beyond gathering known GHS classifications and information on regulatory and M/RSLs, you should obtain comprehensive CHAs on all chemicals, including residuals or impurities above a concentration threshold. Critical data gaps should be filled with information from NAMs as feasible, and testing as necessary. CHAs should include information on the persistence, bioaccumulation potential, and mobility of the chemical in the environment, as well as environmental transformation products.

This information can facilitate better understanding of how the chemical behaves in the environment and how it can be used in a circular economy.

3.7.2 Carbon

By "carbon" we refer to both the source and amount of energy consumed across the product life cycle. While reducing the amount of energy consumed has universal appeal, agreement on the source of energy is a different story. Different energy sources matter because they have different attributes. As mentioned earlier, attributes of energy sources should be confirmed with impact measurement using LCA. There is also social value in shifting to use of renewable energy sources.

At a minimum, you should know both the amount of energy consumed and the source(s) of energy used in your process. Some LCA databases include impact data that serve as proxies for different types of energy sources ranging from renewables such as wind, hydroelectric, and solar to the combustion of coal, natural gas, or other petroleum-based products.

More aspirational actions require more data and expand the consideration of energy consumption and energy sources to the entire supply chain. It can be challenging to know just where a facility's energy comes from. The US EPA's Emissions & Generation Resource Integrated Database (e-GRID) can help to pinpoint the source of energy by location [53]. Environmental product declarations are data intensive (and subsequently expensive), but they provide a more accurate inventory of energy consumed and the associated human health and environmental impacts based on where and how the energy was generated [54]. Traditional LCAs based on software tools like Gabi or SimaPro are commonly used for assessing and comparing products based on carbon [55, 56].

3.7.3 Circularity

By "circularity" we refer to the design of products and processes that decouple economic activity from the consumption of finite resources and eliminate the generation of waste. At the material level, entrepreneurs targeting circularity should focus on dematerialization, re-materialization, and design for value recovery, also called "material re-utilization" [31].

The goal of dematerialization is to use less material and subsequently to reduce impacts by using fewer natural resources. For example, Method Home innovated its laundry detergent by creating a 4x concentrate such that a 900 mL container could provide 75 washes [57].

The goal of re-materialization is to transition from materials derived from natural resources that have high negative impacts to human health and the environment to those that have lower impacts. For example, entrepreneurs can innovate using abundant waste materials such as crab shell or other crustacean fishing waste to create products that can displace materials of concern in needed services. Chitosan is a high molecular weight biopolymer with hundreds of applications in water treatment, textiles, and agriculture, with the potential to replace toxic antimicrobials, inorganic flocculants and coagulants, and pesticides and insecticides [58].

By value recovery or material re-utilization, we refer to the use of materials that can be readily recovered from products at their end of life, and managed for future use in materials and products of equivalent or better value. For example, DSM features a product called

Niaga (the word "again" written backwards), a synthetic polyester carpet used with an innovative adhesive that can be recycled again and again [59]. DSM's design strategy is rooted in simplicity and safety, using fewer materials and only those that have been evaluated and determined to be inherently safe.

Innovation in recycling is key to facilitating a circular economy based on material reutilization. As an example, Carbios is building an industrial demonstration plant for the enzymatic recycling of PET plastic, producing high-value monomers from the process [60]. For nonreusable or nonrecyclable products, such as chemicals discharged to waste water, chemicals should be selected and verified for rapid biodegradability in the medium into which they are discharged. By contrast, durability can be a strong option for some products. For example, LEGO® bricks are made from plastic, but they are durable; LEGO® bricks made 50 years ago fit with LEGO® bricks made today [61].

At a minimum, you should consider the entire life cycle of your chemical technology and "design with the end in mind" – having a plan for end-of-life management of the product and its packaging. You should seek to minimize waste in production and avoid unnecessary packaging, using strategies of dematerialization, rematerialization, and reutilization, in addition to selecting materials that are commonly collected and recycled.

More aspirational actions include detailed and verified waste management plans and directions for users. The How2Recycle program can provide guidance on effective and standardized labeling, and has published a How2Recycle Guide to Recyclability [62]. You could consider product takeback options as part of a product stewardship plan. For example, HP includes a prepaid and pre-addressed envelope with new ink cartridges. Customers can mail empty ink cartridges back to HP at no cost and with minimal effort, where they are reused and recycled into new ink cartridges [63].

3.8 Introducing Your Sustainable Chemical Technology into the Marketplace

3.8.1 Communicating Cost Versus Life-Cycle Benefits

Sustainable chemical technologies may have life-cycle benefits that are not reflected in up-front costs. You should look at the life-cycle costs of your chemical technology and communicate the benefits. For example, in the boat antifouling sector, there can be a broad range in cost per gallon of coatings. Some of the coatings are biocidal, designed to kill anything that attaches to the boat. These can last as little as one year and typically no more than three. Newer alternative coatings are hard ceramics that do not contain toxic biocides and are designed to prevent attachment of organisms to the boat, removing them with friction when the boat moves through the water. These ceramic coatings are more expensive per gallon but can last up to 10 years, making them more economical over the longer term. The savings are reflected not only per gallon of coating, but also because the owner does not have to incur the additional costs (and downtime) of hauling a boat out of the water to re-coat it.

3.8.2 Benefiting from a "Green Premium"

In some markets, consumers will pay more for a product that is perceived to have human health and/or environmental benefits. Growth in green cleaning products and organic food markets are often cited as examples of consumer interest and willingness to pay a "green premium" to know that products are free from toxic chemicals. There are a multitude of green certification programs that can help companies avoid the pitfalls of communicating claims such as those that are exaggerated or not substantiated resulting in legal, financial, and reputational repercussions. Third-party certifications like US EPA's Safer Choice can be incredibly valuable to companies that want to market their products based on human and environmental health attributes, demonstrating with credibility that their products contain chemicals that meet Safer Choice requirements and are safer than conventional products [64].

3.8.3 Avoid Greenwashing

The term "greenwashing" is generally attributed to environmentalist Jay Westerveld, whose 1986 essay described how the hotel industry falsely promoted the reuse of towels as part of a broader environmental strategy when, in fact, the initiative was primarily launched as a cost-savings initiative [65]. Greenwashing can be defined as the practice of falsely marketing products or services as environmentally friendly [66]. Engaging in greenwashing, or "eco-exaggeration," can lead to regulatory liability and damage the reputation of your business. Companies in both Canada and the US have been stung for unsubstantiated "green" claims and have paid the price, either in terms of financial penalties or in damage to corporate reputations.

In the US, greenwashing is enforced by the Federal Trade Commission (FTC). To provide businesses with guidance on substantiating environmental claims pertaining to products and services, the FTC issued "Guides for the Use of Environmental Marketing Claims" (commonly known as the "Green Guides"). The FTC first issued the Green Guides in 1992 and most recently updated them in 2012 [67]. These guidelines provide general principles that apply to environmental marketing claims, insight into how consumers are likely to interpret particular claims, and how marketers can substantiate these claims to avoid deceiving consumers [68]. As an example of financial penalties associated with greenwashing, four major retailers in the US were fined penalties totaling $1.3 million in 2015 because they falsely labeled rayon textiles as made of bamboo [69]. Because bamboo grows quickly and requires little or no pesticides, consumers assume (incorrectly) that bamboo fabric is better for the environment than other textile fibers. However, transforming bamboo fiber into fabric requires the use of chemicals that are often hazardous to the environment. The end product of this process is a fabric called "rayon" or "viscose," which contains no trace of the bamboo plant.

In Canada, the Competition Bureau of Canada enforces greenwashing violations through the Competition Act, the Consumer Packaging and Labelling Act, and the Textile Labelling Act. To assist businesses, the Competition Bureau released a voluntary guide authored by the Canadian Standards Association that identifies best practices for making environmental claims in Canada [70]. Notably, the Competition Act discourages use of vague, nonspecific, incomplete, or irrelevant claims that cannot be supported through verifiable test methods. A recent example of greenwashing in Canada is the advertising of bamboo fabric as an environmentally responsible choice. The Competition Bureau warned Canadian consumers of this type of greenwashing, which has plagued both the US and Canadian markets [71].

The Competition Bureau of Canada provided the following environmental claims recommendations for businesses selling in Canada [72]:

- Claims should not be misleading or likely to result in misinterpretation.
- Claims should be accurate and specific: claims that broadly imply that a product is environmentally beneficial or benign must be accompanied by a statement that provides support.
- Claims should be substantiated and verifiable: claims must be tested, and all tests must be scientifically sound, conducted in good faith and documented.
- Claims should be relevant: claims must be specific to a particular product and used only in an appropriate context. Claims must also take into consideration all relevant aspects of the product's whole life cycle.
- Claims should not imply that a product is endorsed by a third-party organization when it is not.

3.9 Conclusions

For a chemical technology to be sustainable, it must not only have market value and value in use, but it must also confer benefits to human well-being and the environment. Sustainability is a complex, systems-based concept that is not as easy to measure as monetary value. It is important nonetheless to define sustainability for your chemical technology, to set aspirational goals, and to measure progress toward achieving those goals.

As you develop and commercialize your technology, you should clearly understand your design goals and how they align with the UN Sustainable Development Goals. Your innovation should be evaluated across its entire life cycle. Because sustainability is so broad, a useful way to approach the problem as a system is to consider (i) chemistry, (ii) carbon, and (iii) circularity.

Taking a life cycle perspective and using available tools and approaches to benchmark products for chemistry, carbon, and circularity reflects current best practices. This approach will help you understand the sustainability benefits of your innovation and also any negative trade-offs. Doing so positions you to make claims that stand out to consumers and purchasers who prioritize health and the environment and who may be willing to pay a "green premium." You should ensure that claims are accurate and specific and that they are tied to clearly articulated sustainable design goals.

In the closing chapter of *Design for the Real World*, Victor Papanek made the following observation: "Design, if it is to be ecologically responsible and socially responsive, must be revolutionary and radical" [16]. Embracing instead of shying away from such challenges and responsibilities will differentiate your product or service from that of your competitors.

References

1 Bridgwater, W. and Sherwood, E.J. (1956). Value. In: *The Columbia Encyclopedia*. New York: Columbia University Press. 2054.

2 Sauter, D.W., Taoufik, M., and Boisson, C. (2017). Polyolefins, a success story. *Polymers (Basel)* 9 (6): 185. https://www.ncbi.nlm.nih.gov/pmc/articles/PMC6432472 (accessed 29 October 2020).

3 National Oceanic and Atmospheric Administration (NOAA). (2012). Garbage patches. Marine Debris Program. Office of Response and Restoration. https://marinedebris.noaa.gov/info/patch.html (accessed 29 October 2020).

4 UNESCO. (2020). Facts and figures on marine pollution. http://www.unesco.org/new/en/natural-sciences/ioc-oceans/focus-areas/rio-20-ocean/blueprint-for-the-future-we-want/marine-pollution/facts-and-figures-on-marine-pollution (accessed 29 October 2020).

5 National Oceanic and Atmospheric Administration (NOAA). (2020). Ocean pollution. https://www.noaa.gov/education/resource-collections/ocean-coasts/ocean-pollution (accessed 29 October 2020).

6 The Ocean Clean Up. (2020). The great Pacific garbage patch. https://theoceancleanup.com/great-pacific-garbage-patch (accessed 29 October 2020).

7 UN News. (2015). Sustainable development goals to kick in with start of new year. https://news.un.org/en/story/2015/12/519172-sustainable-development-goals-kick-start-new-year (accessed 29 October 2020).

8 United Nations. (2015). Transforming our world: the 2030 agenda for sustainable development. https://www.un.org/ga/search/view_doc.asp?symbol=A/RES/70/1&Lang=E (accessed 29 October 2020).

9 Naidoo, R., and Fisher B. (2020). Reset sustainable development goals for a pandemic world. https://www.nature.com/articles/d41586-020-01999-x? (accessed 29 October 2020).

10 United Nations Global Compact (UNGC). (2020). The SDGs explained for business. https://www.unglobalcompact.org/sdgs/about (accessed 29 October 2020).

11 Wordclouds.com. (2020). WordCloud generator tool. https://www.wordclouds.com (accessed 29 October 2020).

12 Environmental Monitoring Solutions. (2016). How can i do an aspects and impacts assessment? https://www.em-solutions.co.uk/insights/how-can-i-do-an-aspects-and-impacts-assessment (accessed 29 October 2020).

13 Franklin Associates for the Oregon Department of Environmental Quality, Materials Management Section. (2018). The significance of environmental attributes as indicators of the life cycle environmental impacts of packaging and food service ware. https://www.oregon.gov/deq/FilterDocs/MaterialAttributes.pdf (accessed 29 October 2020).

14 Oregon Department of Environmental Quality (2018). Popular packaging attributes. https://www.oregon.gov/deq/mm/production/Pages/Materials-Attributes.aspx (accessed 29 October 2020).

15 Oregon Department of Environmental Quality (DEQ). (2018). Technical summaries. https://www.oregon.gov/deq/FilterDocs/biobased.pdf (accessed 29 October 2020).

16 Papanek, V. (1971). *Design for the Real World: Human Ecology and Social Change*. Knopf Publishing Group.

17 Baum Hedlund. (2020). What is BPA? https://www.baumhedlundlaw.com/consumer-class-actions/bpa-injuries/what-is-bpa (accessed 29 October 2020).

18 United States Environmental Protection Agency. (2015). Bisphenol A alternatives in thermal paper. https://www.epa.gov/sites/production/files/2015-08/documents/bpa_final .pdf (accessed 29 October 2020).

19 DeVito, S.C. and Garrett, R.L. (1996). *Designing Safer Chemicals. Green Chemistry for Pollution Prevention*. American Chemical Society.

20 Radermacher, R. and Hwang, Y. (2005). Introduction. Chapter One. In: *Vapor Compression Heat Pumps with Refrigerant Mixtures*, 1–16. CRC Press.

21 Giunta, C.J. (2006). Thomas Midgley, Jr. and the invention of Chlorofluorocarbon Refrigerants: It Ain't Necessarily So. *Bull. Hist. Chem.* 31 (2): 66–74. http://acshist.scs.illinois .edu/bulletin_open_access/v31-2/v31-2%20p66-74.pdf (accessed 29 October 2020).

22 United States Environmental Protection Agency. (2020). Refrigerant safety: significant new alternatives policy. https://www.epa.gov/snap/refrigerant-safety (accessed 29 October 2020).

23 Midgley, T. and Henne, A.L. (1930). Organic fluorides as refrigerants. *Industrial & Engineering Chemistry* 22 (5): 542–545. https://pubs.acs.org/doi/pdf/10.1021/ie50245a031 (accessed 29 October 2020).

24 Hamilton, J.M. (1963). *The Organic Fluorochemicals Industry*, Advances in Fluorine Chemistry, vol. 3 (eds. M. Stacey, J.C. Tatlow and A.G. Sharpe), 117–180. London: Butterworths. https://www.ag.state.mn.us/Office/Cases/3M/docs/PTX/PTX3020.pdf (accessed 29 October 2020).

25 Molina, M.J. and Rowland, F.S. (1974). Stratospheric sink for chlorofluoromethanes: chlorine atom-catalysed destruction of ozone. *Nature* 249: 810–812. https://www.nature .com/articles/249810a0 (accessed 29 October 2020).

26 Farman, J.C., Gardiner, B.G., and Shanklin, J.D. (1985). Large losses of total ozone in Antartica reveal seasonable ClOx/NOx interaction. *Nature*: 315, 207–210. https://www .nature.com/articles/315207a0 (accessed 29 October 2020).

27 Mulvihill, M.J., Beach, E.S., Zimmerman, J.B., and Anastas, P.T. (2011). Green Chemistry and Green Engeinering: A Framework for Sustainable Technology Development. *Annual Review of Environment and Resources* 36: 271–293. https://www.annualreviews .org/doi/full/10.1146/annurev-environ-032009-095500 (Accessed 9 November 2020).

28 Hogue, C. (2019). Differentiating between green chemistry and sustainable chemistry in Congress. https://cen.acs.org/environment/green-chemistry/Differentiating-between-green-chemistry-sustainable/97/web/2019/07 (accessed 29 October 2020).

29 Anastas, P.T. (ed.) (Vol. eds. Constable, D.J.C and Jimenez-Gonzales, C.). (2018). *Green Metrics*, vol. 11. Weinheim Germany: Wiley-VCH.

30 OECD. (2020). Green growth and sustainable development. http://www.oecd.org/ greengrowth/

31 OECD (2012). Sustainable materials management: making better use of resources. http://www.oecd.org/env/waste/smm-makingbetteruseofresources.htm (accessed 29 October 2020).

32 Ellen Macarthur Foundation. (2017). Concept: what is a circular economy? https://www .ellenmacarthurfoundation.org/circular-economy/concept (accessed 29 October 2020).

33 Fiksel, J. (Aug 2006). A framework for sustainable materials management. *JOM*.

34 Common Objective (2018). Fibre briefing: leather. https://www.commonobjective.co/ article/fibre-briefing-leather (accessed 29 October 2020).

35 Bolt Threads. (2020r). Bolt technology – meet MYLO. https://boltthreads.com/technology/mylo (accessed 29 October 2020).

36 Hirschler, M. (2014). Flame retardants: background and effectiveness. https://www.sfpe.org/page/2014_Q3_3 (accessed 29 October 2020).

37 Interstate Chemicals Clearinghouse. (2017). Alternatives and assessment guide V 1.1. http://theic2.org/article/download-pdf/file_name/IC2_AA_Guide_Version_1.1.pdf (accessed 29 October 2020).

38 ChemSafetyPRO. (2019). Overview of restricted substances lists of big multinational companies. https://www.chemsafetypro.com/Topics/Restriction/Overview_of_Restricted_Substances_Lists_of_Big_Multinational_Companies.html (accessed 29 October 2020).

39 ChemSafetyPro. (2016). GHS and hazard communication. https://www.chemsafetypro.com/Topics/Category/GHS_SDS_label.html (accessed 29 October 2020).

40 Healthy Building Network. (2019). Search Pharos. https://pharosproject.net (accessed 29 October 2020).

41 UNECE. (n.d.). About the GHS: globally harmonized system of classification and labelling of chemicals (GHS). https://www.unece.org/trans/danger/publi/ghs/ghs_welcome_e.html (accessed 29 October 2020).

42 Cradle to Cradle Products Innovation Institute. (2020). Material health assessment methodology. https://www.c2ccertified.org/resources/detail/material_assessment_methodology (accessed 29 October 2020).

43 Clean Production Action. (2018). GreenScreen for safer chemicals hazard assessment guide. Version 1.4. https://www.cleanproduction.org/images (accessed 9 November 2020).

44 Design for the Environment (DfE) Program. (2011). Alternatives assessment criteria for hazard evaluation. https://www.epa.gov/saferchoice/alternatives-assessment-criteria-hazard-evaluation (accessed 29 October 2020).

45 ChemFORWARD. (n.d.). Know Better Chemistry. www.chemforward.org (accessed 29 October 2020).

46 Interstate Chemicals Clearinghouse. (2014). Chemical hazard assessment database. http://theic2.org/hazard-assessment#gsc.tab=0 (accessed 29 October 2020).

47 ToxServices. (2020). ToxFMD Screened Chemistry® library. https://database.toxservices.com/ (accessed 29 October 2020).

48 Verisk 3E. (2020). www.verisk3e.com (accessed 29 October 2020).

49 SciVera. (2020). www.scivera.com (accessed 29 October 2020).

50 National Center for Healthy Housing. (2007). Fact sheet: toys and lead exposure. https://nchh.org/resource-library/Fact_Sheet_Lead_In_Toys.pdf (accessed 29 October 2020).

51 Canadian Centre for Occupational Health and Safety. (2020). Safety data sheet compliance tool. http://whmis.org/sds (accessed 29 October 2020).

52 Society for Chemical Hazard Communication. (2020). SCHC HazCom consultants. https://www.schc.org/hazcom-consultants-main (accessed 29 October 2020).

53 EPA. (2020). Emissions & Generation Resource Integrated Database (e-GRID). https://www.epa.gov/energy/emissions-generation-resource-integrated-database-egrid (accessed 29 October 2020).

54 EPD International. (2020). The International EPD System. https://www.environdec.com/What-is-an-EPD (accessed 29 October 2020).

55 GaBi Software. (2020). Life cycle assessment. http://www.gabi-software.com/america/topics/life-cycle-assessment-lca (accessed 29 October 2020).

56 SimaPro. (2020). SimaPro life cycle assessment (LCA) software. https://simapro.com (accessed 29 October 2020).

57 Method Home. (2020). Method Home laundry products. https://methodhome.com/product-category/laundry (accessed 29 October 2020).

58 TidalVision. (2020). Chitosan. https://tidalvisionusa.com/chitosan (accessed 29 October 2020).

59 DSM-Niaga. (2020). https://www.dsm-niaga.com/contact.html (accessed 29 October 2020).

60 Carbios. (2020). Carbios begins construction on industrial demonstration plant in final step to commercializing its PET recycling technology. https://carbios.fr/en/carbios-begins-construction-on-industrial-demonstration-plant-in-final-step-to-commercializing-its-pet-recycling-technology (accessed 29 October 2020).

61 Cordon, D. and Mattson, C. (2018). The LEGO Brick: The BYU design review. https://www.designreview.byu.edu/collections/the-lego-brick (accessed 29 October 2020).

62 How2Recycle. (2020). How2Recycle guide to recyclability. https://how2recycle.info/guide (accessed 29 October 2020).

63 Knerl, L. (2019). How to recycle toner cartridges responsibly. https://store.hp.com/us/en/tech-takes/recycle-toner-cartridges (accessed 29 October 2020).

64 Rich, E. (2017). Third-party certifications help companies avoid pitfalls in green marketing. https://cleangredients.org/third-party-certifications-help-companies-avoid-pitfalls-in-green-marketing (accessed 29 October 2020).

65 Sullivan, J. (2009). Greenwashing gets his goat. *Times Harold-Record* (1 August). https://www.recordonline.com/article/20090801/NEWS/908010329 (accessed 29 October 2020).

66 Kolluru, S. (2019). How is greenwashing regulated in Canada. *The Varsity*. 21 September. https://thevarsity.ca/2019/09/21/how-is-greenwashing-regulated-in-canada (accessed 29 October 2020).

67 Federal Trade Commission (FTC). (2012). Guides for the use of environmental marketing claims ("green guides"). https://www.ftc.gov/policy/federal-register-notices/guides-use-environmental-marketing-claims-green-guides (accessed 29 October 2020).

68 Federal Trade Commission (FTC). (2020). Environmentally friendly products: FTC's green guides. https://www.ftc.gov/news-events/media-resources/truth-advertising/green-guides (accessed 29 October 2020).

69 Federal Trade Commission (FTC). (2015). Nordstrom, Bed Bath & Beyond, http://Backcountry.com, and J.C. Penney to pay penalties totaling $1.3 million for falsely labeling rayon textiles as made of "bamboo". https://www.ftc.gov/enforcement/cases-proceedings/152-3030/nordstrom-inc (accessed 29 October 2020).

70 Canadian Standards Association (CSA). (2008). Environmental claims: a guide for industry and advertisers. Developed in partnership with Competition Bureau Canada. https://www.competitionbureau.gc.ca/eic/site/cb-bc.nsf/vwapj/guide-for-industry-and-advertisers-en.pdf/$FILE/guide-for-industry-and-advertisers-en.pdf (accessed 29 October 2020).

71 Competition Bureau of Canada. (2019). Don't be bamboozled: the real deal with "bamboo fabric". https://www.canada.ca/en/competition-bureau/news/2019/02/dont-be-bamboozled-the-real-deal-with-bamboo-fabric.html (accessed 29 October 2020).

72 Competition Bureau. (2017). It's not easy being green. Business must back up their words. https://www.canada.ca/en/competition-bureau/news/2017/01/not-easy-being-green-businesses-must-back-up-their-words.html (accessed 29 October 2020).

4

Intellectual Property Management and Strategy
Nick Sutcliffe

Mewburn Ellis LLP, Cambridge, UK

4.1   Intellectual Property

The concept of property is a cornerstone of developed societies. People own things, and the ownership of a thing confers certain legal rights over that thing. The legal rights conferred by the ownership of property are enshrined in laws that, for example, make it illegal to steal property owned by someone else [1].

The things that are owned by people are generally physical objects that can been seen and touched, like a tennis racket, car, or coffee machine. Because these objects are tangible and exist in physical form, everyone can see not just that the thing really does exist, but also exactly what the thing is. For example, it is immediately evident from picking up a tennis racket not just that the racket exists, but also what its size, shape, and other physical properties might be. The ownership of a physical object may also be evident from its location or marking. For example, a tennis racket may be labeled with the owner's name or kept in the owner's locker.

Intangible things, such as ideas or information, which are not physical objects, are more of a problem. Because you cannot see or touch something that is intangible, it may not be evident exactly what or where the intangible thing is or even whether it exists at all. Given these innate uncertainties, how can you know if you own something that is intangible, or even precisely what the intangible thing is that you own?

The ownership of intangible things is the realm of intellectual property rights (IPRs). An IPR is an item of property that consists of a set of legal rights over an intangible thing [2]. To be protected by an IPR, the intangible thing must be defined so it is clear exactly what is covered by the legal rights. As well as allowing someone to possess ownership rights over an intangible thing, IPRs therefore also define what the intangible thing is.

This is relevant because a new technology in its raw form consists of a mixture of intangible ideas, information, and experimental data. An IPR, such as a patent, may not only crystalize a definition of the new technology from all these ideas, information, and data, but it also confers ownership rights to the technology that is defined. The ownership rights to a new chemical technology that are conferred by an IPR are likely to be key elements in commercializing the technology. Indeed, the IPRs that confer ownership of a new technology

How to Commercialize Chemical Technologies for a Sustainable Future, First Edition.
Edited by Timothy J. Clark and Andrew S. Pasternak.

may represent the bulk of the value of an early-stage technology business and so are key assets of the business.

4.2   What Is an Intellectual Property Right?

IPRs turn the science that underlies new chemical technologies into commercial assets. These allow the technologies to be developed into products and processes that can be brought to market.

IPR is an umbrella term that covers a number of different rights, including copyright, designs, trademarks, and patents. Each of these rights protects a different type of intangible asset. For example, copyright protects the expression of ideas and concepts. In addition to literary or artistic works, copyright may also be relevant to documents, design drawings, and software that is generated by a technology business. Designs protect the appearance and visual form of a product, while trademarks protect the brand names and logos associated with a business. Some of these rights arise only after an application process, while other rights arise automatically, without any action being needed. More niche types of IPRs exist in many countries and may be appropriate to certain businesses including database rights, plant variety rights, integrated circuit topography rights, and vessel hull design rights. However, these rights are unlikely to trouble most businesses seeking to commercialize technologies in the chemical space.

For a sustainable chemical technology business, the key form of IPR is likely to be the patent. Patents protect inventions in technical fields and are likely to be the focus for investors and other parties looking to commercialize a technology. Patents are therefore key assets for a technology business. A business may also look to trade secrets and know how to protect valuable innovations for which patent protection is either not available or not desirable. Trade secrets can be legally enforceable [3], and trade secret laws have recently been harmonized across the EU [4] with a view to providing a comparable level of protection to other major jurisdictions, such as the US [5].

4.3   The Value of Intellectual Property Rights to a Sustainable Chemical Technology Company

Whether an inventor, university, start-up, or established business, anyone who is attempting to commercialize a new technology has many demands on their time and resources. However, since IPRs are likely to represent key assets, time and attention devoted to IPRs from the outset are likely to be well spent.

IPRs that cover a technology allow the owner to stop competitors using the technology. Indeed, would-be competitors may observe a well-managed portfolio of IPRs covering a technology and look for alternative approaches that are not so well-protected. This ability to exclude other people from using a technology can make IPRs valuable. In the absence of competition, a business can charge a premium price and make more profit. If the technology is genuinely revolutionary or groundbreaking, the monopoly provided by solid IPRs can be extremely valuable.

IPRs are recognized by investors and potential partners as an asset. Investment in an early-stage business that is trying to commercialize a new technology may be contingent on the existence of strong IPRs covering the technology. IPRs provide some reassurance that it will be possible to make a return on any investment without competitors copying the technology and eating into potential profits. A portfolio of well-managed IPRs is likely to significantly increase the attractiveness of a business to investors.

Conversely, if the business does not own any IPRs covering its technology, then competitors may be free to adopt the technology if it turns out to be successful. This may deter investment in the business since returns may be small or nonexistent if the technology is commercialized.

The development and commercialization of a sustainable chemical technology often involves collaborations with other parties. IPRs may provide a date-stamped definition of the technology that is brought into a collaboration. This defines the technologies that each of the parties brings into the collaboration and may be useful in preventing disputes later about exactly what technology is owned by each of the parties.

IPRs may also be a source of revenue in themselves. Since they are commercial assets, IPRs may be sold to third parties. For example, a business wanting to focus on a core technology may sell IPRs that relate to other technologies. As an alternative to selling an IPR directly, revenue can also be generated by granting a license.

Owning IPRs that cover a new technology gives the owner the right to stop other people from using it. Crucially, *they do not give the owner the right to use the technology*. For example, in some cases, it may not be possible to use a new technology without falling within someone else's existing IPR. If the use of a new technology is blocked by the existence of other people's IPRs, then the technology may be difficult to commercialize. For this reason, it is important to establish "freedom to operate" (FTO) at an early stage and to continually update and check as the technology itself is modified and other people's IPRs are granted. This is likely to be a key issue for investors, who will want to be reassured that the technology they are investing in can be used commercially to generate a return on their investment. If a blocking IPR is identified, then it may be necessary to obtain a license to the blocking IPR or to alter or adapt the new technology to avoid it.

The existence of a well-managed IPR portfolio may also be useful in negotiations with other parties. For example, in situations where both parties have an IP position, it may be possible to resolve an FTO issue with a cross-licensing agreement in which a business receives a license to the IPRs of a third party in return for giving the third party a license to its own IPRs. This sort of resolution is possible only when a business has an IPR portfolio to bring to the negotiating table.

Devoting too much time and attention to IPRs may also be problematic for early-stage companies. First, many IPRs are expensive to obtain and maintain. Merely owning IPRs does not guarantee revenue or income, and the IPRs may simply be a drain on resources. Prudent planning is required to ensure that a business pursues IPRs that are likely to support its current or future commercial activities. Some pragmatism may be needed to avoid costly overstretch in terms of overly speculative patent applications or protection in countries that are not of strategic interest.

IPRs can also consume the time of staff at many different levels in an organization from scientists to CEOs. Again, prudent management may be needed to prevent staff from being sucked into an IP vortex and allow them to actually develop the technology on which the business's future depends.

4.4 Patents Explained

4.4.1 What Sort of Technology Can Be Patented?

Patents can be awarded for inventions in any technology space [6, 7], including all manner of sustainable chemistry and clean technologies. The principles of the patent system remain the same regardless of the technology, although certain patent offices, including the UK [8] and Canada [9], have implemented processes to encourage patent applications in the cleantech space.

4.4.2 What Is a Patent?

A patent is an exclusionary right that allows a patent owner to stop other people from performing certain activities relating to the invention that is defined by the patent [2]. For example, if the invention defined by a patent is a product, such as a new chemical, the owner of the patent can stop third parties from making, using, selling, or importing the new chemical. This is a valuable monopoly that stops people from competing with the patent owner so that the patent owner can charge a premium price for the patented product without fear of being undercut. For this reason, businesses that are investing in the commercialization of a new technology will generally want patent protection for their products to provide a return for their investment in the form of higher profits.

4.4.3 The Patent Bargain

Because it allows patent owners to charge premium prices, the monopoly provided by a patent is potentially detrimental to consumers. However, governments tolerate this anticompetitive effect in order to encourage innovation and creativity. A patent represents a deal between the state and the innovator. The state grants a monopoly to the innovator for the invention that is defined in a patent application. In return, the innovator has to describe how to perform the invention in detail in the application document. This means that the invention is made available to the public at large when the patent application is published. This bargain of monopoly in return for disclosure ensures that the technology is made public, and further advances can be made on the back of it. In the absence of a patent system, there would be no incentive to disclose new technologies, and innovators would attempt to gain a competitive edge by keeping their technology secret.

4.4.4 Territorial

A patent is a national right that applies only to a particular country. Outside that country, a patent has no effect. For example, activities in the UK will not infringe a Canadian patent, or vice versa. There is no such thing as a worldwide patent, and separate patents must be applied for in all the countries in which protection is required.

Applying for a patent in every country in the world would be prohibitively expensive, so a patent applicant needs to select the countries in which to seek (or not seek) a patent to protect an invention. Fortunately, various international treaties, notably the Paris Convention [10] and the Patent Cooperation Treaty (PCT) [11], allow the decision about the countries in which to pursue patent protection to be delayed for some time after the initial filing. Thus, a complete filing strategy is not required at the outset of the application process.

4.4.5 Time Limitation

Patents are not only limited by territory, but also by time and consequently expire after a certain date (see Figure 4.1). After a patent has expired, anyone is free to use the invention disclosed in the patent. In general, patents expire 20 years after the initial patent application was filed. However, there are certain exceptions to this principle. For example, additional term (time) may be granted to patents on inventions that are subject to a regulatory process, such as pharmaceuticals. The additional term is provided by a Supplementary Protection Certificate in Europe [12] and a Patent Term Extension in the US [13]. The US will also grant additional patent term to compensate for delays caused by the US Patent Office during the application process [14].

4.4.6 Property

Patents (and applications for patents) are items of property that can be bought or sold. This is important because the route to market of a new technology often involves trading or licensing the rights to the technology to other parties that are better placed to further develop and commercialize it. Transferring or licensing patents and other IPRs generally requires a formal written agreement, which can be lengthy and highly detailed. Assignments and licenses of IPRs may generate revenue for the owner of the IPR and are often a route by which IPR owners (such as universities) transfer the rights to a new technology to entrepreneurs to commercialize it.

4.4.7 Exclusionary Right

A patent is an exclusionary right that does not provide the patent owner with any positive right to use the invention that is defined by the patent. This may have important consequences for the patent owner when an invention cannot be put into practice without falling within the scope of a third-party patent. The importance of identifying and navigating around third-party patents is discussed in more detail later.

4.4.8 Criteria for Patentability

To be valid, the invention defined in a patent must meet certain criteria. In some countries, patents are granted on all applications, and these criteria are considered only if there is a dispute involving the granted patent. However, in all major countries, these criteria are assessed during the application process and a patent is granted only if they are considered to have been fulfilled.

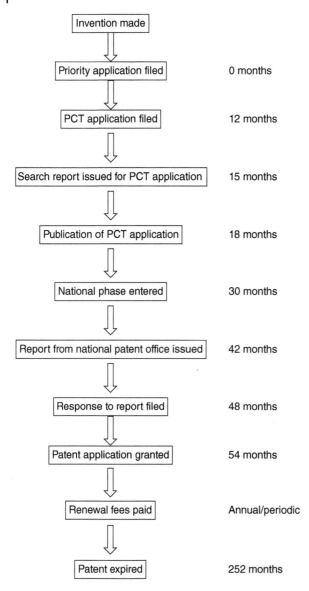

Invention made	
Priority application filed	0 months
PCT application filed	12 months
Search report issued for PCT application	15 months
Publication of PCT application	18 months
National phase entered	30 months
Report from national patent office issued	42 months
Response to report filed	48 months
Patent application granted	54 months
Renewal fees paid	Annual/periodic
Patent expired	252 months

Figure 4.1 General timeline for the patent process. Most time points are firm, while others are estimates.

First, to be awarded a patent, the invention described in a patent application must be the right sort of invention. This means that patents will not be granted for nontechnical inventions, such as aesthetic creations, rules for playing games, or a host of other types of inventions which lawmakers do not consider appropriate to grant patents [15, 16]. Precisely which types of inventions will not be awarded patents differs from country to country. In particular, patent law in the US has recently become complicated and confusing as to which inventions are eligible for patent protection and which are not. This particularly affects diagnostics [17] and computer methods [18]. However, most chemical technologies will be considered patent eligible in most countries, so this criterion is unlikely to significantly impact patents in the sustainable chemistry space.

An invention must also be novel to be awarded a patent [19, 20]. This means that **the invention must not have been disclosed to the public by any means anywhere** before the date on which the patent application was filed. This can present challenges for IP professionals in ensuring that inventions are protected by patent applications before the inventors submit scientific papers for publication or the CEO discusses their team's latest findings at a conference. In many countries, a "grace period" exists in which a patent application can be safely filed *after* the disclosure of an invention [21, 22]. Disclosures before filing in these countries may not impact on the subsequent patent application. However, in many countries including major regions such as Europe, a disclosure of the invention before a patent application is filed will likely be absolutely fatal.

An invention must also be inventive to be awarded a patent. This means that the invention must not be obvious to a notional person familiar with the field, the so-called "person skilled in the art" or "person having ordinary skill in the art" [23, 24]. Whether an invention is "non-obvious" or involves an inventive step is determined using specific legal tests that differ slightly from country to country. In general, if an invention achieves some technical advantage that could not be predicted based on what was already known, then it is likely to be considered inventive. Routine or inconsequential modifications of a known technology are unlikely to be considered inventive.

The patent application must also describe the invention in sufficient detail to allow the notional person familiar with the field to perform the invention [25, 26]. In many cases, this means the application should include experimental data to confirm not only that the invention works, but also that every variation of the invention that falls within the definition of the invention in the application will work (or is likely to work).

These criteria are assessed during the application process by dedicated patent examiners. Patents are awarded only for inventions that are considered to meet the criteria. Preparing a patent application and successfully negotiating it through the application process can be complicated and expensive, and many businesses rely on specialist agents or lawyers to handle their patent applications and steer them through the process. Figure 4.1 shows a simplified schematic of the application process.

4.4.9 Preparing and Filing a Patent Application

The first step in the application process is the preparation of the patent application document. The document needs to include precise definitions of the invention for which a patent is requested, as well as a detailed description of the invention itself. The requirements of setting a date for the invention mean that it is important that the application document be as complete as possible from the outset [27]. Typically, the document is drafted in collaboration with a patent attorney/agent. Once finalized, the application document is filed at a patent office.

The date on which the patent application is filed at the patent office is significant. This becomes the date of the public knowledge (the "prior art") against which the novelty and inventiveness of the invention in the patent application will be assessed [28]. Anything made available to the public before the filing date can potentially destroy the novelty or inventiveness of the invention in the patent application. Anything made available to the public on or after this date is not generally considered. This means that you can talk more freely about your invention without endangering your patent application at this point.

4.4.10   12-Month Anniversary

After it is filed, nothing generally happens to the patent application for 12 months. The period is known as the "priority period" or "priority year" [29]. During the priority year, it is possible to file further patent applications for the same invention that are entitled to the filing date of the original application (the "priority application"). These further applications provide an opportunity to include additional experimental data, modifications, or variations that were not available or described when the original application was filed. In addition, the priority year also provides an opportunity to file further patent applications for the invention in other countries.

The priority period means that the first anniversary of the date of the original patent filing is an important decision point in the application process. After this point, it is no longer possible to add further data or other information to the patent application. Nor is it possible to extend the patent application to countries beyond those envisaged at the first anniversary.

The date of the first anniversary of the filing of the original patent application should therefore be carefully monitored so that actions can be carried out and decisions made in good time. The original patent application needs to be reviewed carefully to identify whether any obtained experimental data shows further development of the invention since the original filing or whether new modifications or variations have been developed that now need to be described in the application. For example, the original application may be focused on a specific compound that was found to have a useful activity, but subsequent experiments may have found that other compounds have the same effect. A further patent application might include all of these new compounds and may also define a class of compounds that shows the effect.

The business also needs to think carefully about the countries in which the patent application might be pursued. One option is to file further patent applications at the 12-month stage in all of the countries in which protection for the invention is needed, with all of these further patent applications claiming the "priority," and hence the filing date, of the original patent application. The downside of this approach is the cost of pursuing parallel patent applications in different countries at an early stage. However, it might be a suitable strategy when patent protection is required quickly in only a few countries.

After filing, each parallel patent will then be published by the respective national patent offices around 18 months from the date of the filing of the original patent application. The patent application will then be searched. This means a patent examiner will review the application and search for scientific papers, patent applications, or other material in the public domain that might disclose (and potentially invalidate) the invention described in the patent application.

4.4.11 PCT Applications

An alternative to filing multiple applications during the 12-month stage is to file a PCT application [30]. This is a single application that keeps options open for patent protection in any of 152 signatory countries, including all the major jurisdictions. The great advantage of this approach is that it delays the cost of filing parallel patent applications in different countries for an additional 18 months. This means decisions about the countries in which to protect an invention do not need to be made until 30 months after the original patent application. This delay may be invaluable in sizing up the commercial opportunities for a technology before committing lots of resources to protect it. In addition, the PCT process provides a preliminary search and examination, which will give a good indication of the prospects of patents being successfully granted, before the decision point is reached.

After filing, a PCT application is subjected to a search by one of the major patent offices [31] (referred to at this stage as the International Search Authority [ISA]). An examiner will issue a search report listing the documents that they have found, along with some comments about the patentability of the invention in light of these documents [32]. Although there is no need to respond to these comments, an optional examination procedure exists within the PCT process to try to persuade the examiner to revise their comments if they are initially negative. This is sometimes useful in influencing other examiners in individual countries later in the application process.

By the end of 30 months from the filing date of the original application, a PCT application must enter the national phase in all of the countries in which a patent is desired. Once entered into the national/regional phase, a patent application is treated by the patent office concerned in a similar manner to a directly filed patent application.

4.4.12 Patent Examiners

While some smaller countries rely heavily on the results of examination elsewhere, most large countries have their own patent office These contain a cohort of patent examiners who assess patent applications according to the specific criteria of that country and decide whether to grant or refuse them. Patent examiners are also responsible for searching the published literature to find material that might invalidate a patent application. Patent examiners specialize in searching and examining patent applications. They invariably have a technical background and may have a doctorate and often post-doctoral experience. In large patent offices, such as the EPO or USPTO, examiners tend to work in tightly defined technical fields and so are likely to have experience and expertise in the subject matter that they examine. In smaller patent offices, examiners may work across larger technical areas and have less familiarity with the technology.

4.4.13 Patent Examination

Whether filed directly at the patent office of a country or region or indirectly via the PCT system, a patent application is subjected to examination in most countries by a patent examiner to see whether it meets the criteria for patentability. Typically, examination involves the issuance of at least one report by the examiner explaining why the invention defined in the patent application does not fulfill the criteria for patentability and at least one response by the applicant making changes to the definition of the invention and/or explaining why the criteria are fulfilled.

Eventually, this process leads to either grant of a patent or refusal by the examiner. If the application is refused, most patent offices have appeal processes that can be initiated to challenge the refusal.

4.4.14 Grant

Once a patent is granted on an application, the business has an enforceable right that can exclude other people from using the invention in that jurisdiction.

4.4.15 Renewal Fees

Renewal fees are official fees that are paid to a patent office. In some countries, renewals are paid only on granted patents. In other countries, renewal fees must be paid even when the application is pending. Renewals are usually paid annually, although the US is a notable exception in which maintenance fees are only paid $3\frac{1}{2}$, $7\frac{1}{2}$, and $11\frac{1}{2}$ years after grant [33].

4.4.16 Costs

Engaging in the patent application process is not cheap. Not only will you need to pay the fees of your patent attorney or agent but also a whole range of official fees, translation costs, renewal fees, and foreign attorney charges. Managing all these costs and balancing them with the other needs of the business is an important aspect of any patent strategy.

For a typical invention in the sustainable chemistry field, the preparation of the initial priority application might cost $6500–$13 000 (£5000–£10 000), which will mostly be the fees of your attorney or agent. After the filing of this priority application, costs for the next 12 months are likely to be low.

At the 12-month stage, action will need to be taken and therefore costs incurred. Typically, a PCT application is filed at this stage. This might cost about $5000–$8000 (£4000–£6000). Unless there are major revisions of the text, these charges will mostly be made up of official fees. The costs for the next 18 months are unlikely to be substantial and will cover the costs of correspondence with your attorney or agent and discussions about prosecution strategy.

The 30-month stage is a major cost point with charges of $5000–$6500 (£4000–£5000) for each country in which you initiate the national phase of your PCT application. This is made up of a combination of attorney/agent fees, charges for foreign agents, and official fees. In non-English-speaking countries, a translation may be required, and this is likely to cost an additional $4000–$6500 (£3000–£5000), depending on the amount of text that needs translating.

From this point on, you may be prosecuting separate patent applications in parallel foreign countries with all of the associated costs. Although they will vary greatly, depending on the circumstances, you can expect it to cost at least $13 000–$26 000 (£10 000–£20 000) to reach grant (or refusal!) in each country. These costs will be spread over anything from two to three years in rapid countries to 12–14 in slower ones.

After the patent is granted, the only costs will be renewal or maintenance fees. These are payable in each country where a patent is in force, so the total cost of maintaining a patent will depend on the number of countries involved. As a rough estimate, renewing a patent might over 20 years cost a total of about \$4000 (£3000) in Canada, \$13 000 (£10 000) in the US, \$6000 (£4700) in the UK, \$16 000 (£12 000) in Germany, and \$6500 (£5000) in France.

4.5 Building an IP Portfolio

4.5.1 Invention Management

Well-defined internal processes are an important first step in building a solid ring fence of IPRs around a new technology.

Although the importance of well-kept laboratory notebooks has diminished with recent changes in US patent law, they may still be important in documenting when an invention was made, what was done, and who did it. This may be invaluable later in legal proceedings, if there are any disputes about the invention. To be of most use in any legal dispute, traditional laboratory notebooks should be permanently bound and have consecutively numbered pages. In addition, entries in the notebooks should be recorded in permanent ink and each page signed and dated by the person performing the experiments and by a corroborating witness. More modern electronic laboratory notebooks (ELNs) may also be used. Suitable ELNs for documenting inventions must be secure, allow corroboration with electronic signatures, and prevent subsequent editing or deletion of the time, authorship, or content of entries. Since the information recorded in ELNs may become important at any time during the life of a patent (or even after its expiry), attention also needs to be paid to the long-term storage and backup of ELNs.

Internal processes should also be in place to control the technical information that goes out of an organization. This may mean reviewing and approving planned publications, presentations, and other disclosures in advance and also putting confidentiality agreements in place with collaborators. Any information within the company that is of commercial value should be identified and steps taken to ensure that it is treated as confidential. This control of information prohibits disclosures that could prevent valuable inventions from being protected by a patent or trade secret.

The experimental data generated by the scientific team needs to be reviewed regularly to identify possible inventions and capture any intellectual property that might arise from it. Scientists often do not appreciate the potential commercial significance of their work and may be keen to present or publish their exciting new data. Passive reliance on the scientific team to assess their experimental results and flag up potential new inventions to the management team is a risky strategy. Reviewing the data generated by the scientific team on a regular basis allows potential new inventions to be identified at an early stage. This enables a strategy to be put in place around these inventions that takes account of any publications and presentations planned by the scientists. It may also direct research projects toward experiments that will provide useful supporting data for a patent application.

4.5.2 Deciding Whether to File a Patent Application

Once a possible invention has been identified based on the experimental data generated by the scientific team, it needs to be assessed. There are likely to be two stages to this assessment.

The first stage is to determine whether the invention has any commercial value. Inventions with commercial value include products that improve on what is currently available and have a large potential market and anticipated sales. However, more modest inventions that improve the efficiency of a production process may also be of commercial value by providing an edge over competitors. If no commercial value of any kind can be identified, then the invention and indeed the work itself may not be worth pursuing any further.

If an invention is considered to have some commercial value, then the second stage of the assessment is whether the invention might meet the criteria for patent protection, in particular whether it is likely to be both novel and inventive. A definitive answer is not possible at this (or indeed any other) stage because you can never be entirely certain that an invalidating prior art disclosure does not exist somewhere. The scientific team members are likely to be familiar with the scientific literature in their field. If they have confidence that the invention has not been reported previously in the literature, then this might be sufficient comfort to justify an initial patent filing. However, some further searching might also be worthwhile. Searching means reviewing the scientific and/or patent literature to see whether the invention has already been described somewhere. If the invention has already been described, then it is unlikely to be patentable and so pursuing a patent application would not be worthwhile.

Searching is generally carried out using online databases. A range of databases exist, some of which are free to access and others are available only by subscription. While simple searches are straightforward to perform on free online databases, performing more complex search strategies (e.g. combining different elements of keyword, classification, and name searching) is a skilled job. The expertise of a specialist search firm may represent a good investment for the entrepreneur in providing reassurance that a search has been thorough, and the risk of overlooking the existence of damaging prior art is minimized [34].

However thorough it may be, no search can be totally guaranteed to find all relevant prior art documents that may exist. There is always some degree of risk that a damaging prior art document will emerge later. However, the more extensive the search, the more confidence is provided that the invention might be patentable if no damaging prior art is found. The downside is that the cost increases as the search becomes more extensive. Inevitably, the search strategy that is performed will be a compromise between comfort and cost. Time constraints may also limit the scope of the search. In some cases, an early-stage company may prefer to perform only modest initial searching before going ahead with a patent application. As the patent application will be searched by one or more patent offices during the patent application process anyway, this approach may be more cost-effective than spending time and money on extensive pre-filing searching.

If an invention looks like it might be patentable, it may be worth consulting a patent attorney/agent at this early stage to get a feel for the breadth of the protection that might be conferred by a patent. If the only patent that is likely to be granted by a patent office has a very narrow scope, then the patent may be easily worked around by a competitor and so the protection that it provides may not be worth the expense of obtaining it. A patent attorney/agent may also be helpful in advising on whether there is sufficient experimental data to support a patent application and what further experimental data might be needed.

4.5.3 Inventions Not Patentable or Worth Patenting

Even if an invention is unlikely to be patentable, it may still have some value. This value can be captured by keeping the invention secret. No registration is required, but prudent businesses should act to ensure that their secrets and know-how are captured and systems put in place to protect them. It may be important down the line that the business is able to demonstrate that it took reasonable steps to keep its confidential information secret. For example, access to information about the invention within the business may be restricted and anyone with access to it made aware of its confidential nature. Workplace computer systems may be monitored to detect and prevent data breaches. Employment contracts may contain specific provisions regarding the secret inventions and information of the business, and the duty to keep this material confidential may be emphasized at induction and exit interviews. This may be especially important in dealing with the scenario of employees with access to the information leaving the company [35]. Compared to other IPRs, trade secrets are cheap to acquire and have an indefinite life span as long as they can be kept secret. Because patent protection is expensive, time-limited, and requires that the invention is revealed to the world when the patent application is published, some types of invention may be better protected as trade secrets. These include inventions that will only be used in-house and are not susceptible to reverse engineering.

4.5.4 Patent Attorneys/Agents

Acquiring and managing IPRs is complicated, and given their potential value, the IPRs of many sustainable chemical technology businesses will be handled by specialist patent professionals. Businesses with large portfolios of IPRs may employ these patent professionals in-house, while other businesses might engage patent professionals from external legal firms.

Patent professionals who are qualified to handle patent applications include patent attorneys and patent agents. In some jurisdictions, these two terms are interchangeable. In others, a patent agent is able to prepare and file patent applications and represent clients in patent office proceedings while a patent attorney is a qualified lawyer who can also practice more broadly in IP matters.

In almost all cases, patent attorneys and agents have a technical background and are likely to have doctorates in science or engineering and/or industrial experience in research and development. While it is unlikely to find a patent attorney or agent who is already an expert in the specific technology of your business, attorneys or agents with a background in the wider field are likely to be available. These attorneys or agents will be able to get up to speed on your specific technology quickly. Indeed, an ability to understand technical issues quickly is one of the key skills of a patent professional.

Choosing the right patent attorney or agent for a business can be difficult. The choice is commonly based on prior contact, recommendation, or internet search. Since the attorney or agent is likely to play an important ongoing role in protecting key assets, it is important to select someone who is a good fit and understands both the technology and the commercial concerns of the business. Factors to consider include whether the attorney is located nearby or far away, their billing practices and, most importantly, whether or not you have a good working relationship with them. IPRs are complicated, and you will invariably have questions for your attorney or agent. If you find that you are reluctant to pick up the phone to them or it is difficult to get through, then it may be time to look elsewhere.

4.5.5 Ownership

Ownership of IPRs is often a key issue for investors, and a business should ensure that the ownership of their inventions is secure and can be clearly demonstrated. A key question when assessing the value of a sustainable chemical technology business is whether there is a chain of entitlement confirming that the business owns all of their IPRs, so it is a good idea to have this paperwork in place and ready for when it is needed.

Ownership of an invention generally begins with the people who made the invention (i.e. the inventors). In some cases, it will be clear who the inventors are, but in other cases a detailed analysis may be needed to establish who made an inventive contribution and should therefore be named as an inventor. If the inventors are not employees of the business, then the invention will need to be transferred from the inventors to the business with a legal assignment document. The position may be more complicated if the inventors are employees of another party, such as a university, and a chain of assignment documents may be needed to transfer ownership to the business. While assignments can be prepared and signed after a patent application is filed, it is best practice to complete assignments as soon as possible, while the inventors are available and willing to sign.

If the inventors are employees of the business and the invention was made as part of their employment, then in most countries the ownership of the invention is likely to automatically belong to the business that employs the inventors. This can be made clear in employment contracts with specific clauses relating to inventions. Even if they are employees, it may also be good practice for the inventors to complete an assignment to ensure that there is an easily demonstrable chain of title.

4.5.6 When to File a Patent Application

Once a business has decided to file a patent application to protect an invention, the next factor to consider is *when* to file the patent application. There may be a hard deadline by which the application must be filed, such as a public disclosure of the invention. For example, if the scientists' are presenting the invention at a conference and cannot be dissuaded, the patent application will need to be filed no later than the same day as the presentation. Getting this wrong means that patent protection is impossible in Europe and many other countries.

There may be pressure to file the patent application quickly. For example, a business may want to file a patent application before it embarks on a funding round or collaboration or because it is concerned that a competitor will file their own patent application for the same invention.

However, in many cases, it may be advisable to resist these pressures as much as possible and delay the filing of a patent application. Experimental data showing that the invention works as claimed is an important part of any patent application. If experimental data is absent or incomplete, there is a risk that the patent application will not be granted or will be granted only in a narrow form. Delaying the filing of the patent application until robust experimental data is available is likely to lead to a much stronger patent. If there are no planned disclosures and no reason to believe competitors are working on similar inventions, resisting the urge to file the patent application until the supporting data is comprehensive may be the best option.

Even if the patent application must be filed before key experimental data is available, the date of the filing may be synchronized with the planned experimental program so that the key data becomes available before the 12-month anniversary of the date of filing. This allows the key data to be included in the PCT or other application filed at the 12-month stage. Additional experimental information obtained after the 12-month anniversary of the date of filing cannot generally be included in the patent application.

4.5.7 Where to File a Patent Application?

It does not really matter where the first patent application is filed in the initial stage of the patent application process. The purpose of this first patent application is simply to acquire a filing date, and this can be achieved by filing the application at the patent office of any major country. International convention means that all major countries will accept the date of a patent application that was filed in a different major country. The choice of where to file the first patent application is usually therefore a matter of convenience and typically a business will file the patent application at the national patent office of the country in which the business is located. Some countries have legal provisions to restrict applicants from filing first patent applications in other countries. Although focused on military technology, these provisions are often broad enough to cover clean technologies, so it may be worth taking legal advice if a first patent application is to be filed outside the applicant's own country.

In principle, the issue of the countries in which to file a patent to protect an invention must be addressed at the 12-month anniversary of the first patent application [29]. In practice, this issue is often deferred by filing a PCT application. This delays the choice of countries for another 18 months (up until 30 months from the date of the first patent application). Since pursuing parallel patent applications in many countries can be extremely expensive, this 18-month delay can be valuable to a business in assessing whether the invention is sufficiently important to justify this expense.

The choice of countries in which to pursue patent applications for an invention is likely to be a balance between cost and protection. Extensive worldwide filing programs are prohibitively expensive, especially for an untested technology. Even very large companies are likely to make strategic selections about where to file. Factors that might inform these selections include the potential size of the market, the location of competitors, and the ease with which any granted patent can be enforced. Given its size and economic development, the US is typically the first choice of most applicants, followed by Europe, China, Japan, Canada, and Australia.

4.5.8 Controlling the Speed of the Process

The pace of the patent application process is slow in most countries, and funereal in others. In the normal course of events it may take years for a patent to be granted on an application. This slow pace suits many sustainable chemical technology businesses, because it means that the cost of pursuing the patent application is spread over a long period. This may be especially useful where the patent application covers a technology that is still being developed and is not yet generating any revenue. However, there may be circumstances in which it would be helpful to achieve a patent more quickly. For example, a granted patent may make a technology more attractive to investors in a funding round, increase the potential value of a licensing deal, or deter a potential competitor.

Simply tightening up prosecution strategy, for example by responding to communications from patent offices quickly and abandoning over-ambitious claims, can have a significant impact on the speed of a patent application. More formal measures are available at many patent offices to increase the speed at which applications are processed. For example, the EPO [36], USPTO [37], and Canadian Patent Office (CPO) all have accelerated examination procedures. The CPO, in particular, offers an accelerated examination program specifically for technologies that "help to resolve or mitigate environmental impacts or to conserve the natural environment and resources," which may be useful in obtaining granted Canadian patents for clean technologies quickly [9].

4.5.9 Managing the Patent Application Process

The process of applying for a patent typically involves a series of deadlines by which the applicant needs to take some action. Sometimes the required action is a formality, such as paying a fee. Other times, the applicant will need to file a paper explaining why the invention defined in the patent application meets the criteria for patentability. Missing a deadline can be fatal and may mean that the patent application is irretrievably lost. Because the consequences of missed deadlines are so dire, proper systems for recording and monitoring deadlines are essential in managing patent portfolios. Sustainable chemical technology businesses that do not have an in-house IP department generally rely on their patent attorneys/agents to keep track of all the deadlines on their patent applications.

A response to a patent office may need a strategic decision by management or technical input from a scientist. It may take time to get these decisions made and assemble the relevant information, so planning is important at an early stage to ensure that the patent attorney/agent has complete instructions for the response well before the deadline. If the deadline can be extended, it is often tempting to do so. However, extending deadlines increases the overall cost of the patent application and the end result is often a further deadline that cannot be extended (with the applicant in no better position to respond than before).

To develop and commercialize a new technology, a business is likely to be involved in rounds of investment as well as partnering and collaborations with other parties. Before devoting time and resources to any relationship, these other parties are likely to undertake due diligence on the business. One issue of high interest will be the patents and other IPRs of the business. This means that the files relating to each patent and patent application should be well-organized and complete so that it is easy for these other parties to check the key documents. Due diligence can arise suddenly and with great time pressure, so it is wise to insist that files are carefully organized from the outset and that they are continuously maintained and updated.

4.6 Avoiding Other People's IPRs

4.6.1 Freedom to Operate

A sustainable chemical technology business may go to great lengths to protect its technology with patents and other IPRs. However, IPRs do not provide any positive right for the business to actually use the protected technology. It may be that the use or implementation of the technology will fall within the scope of an IPR owned by a third party. This can be extremely serious. The existence of blocking IPRs owned by a third party may deter investment in a new technology and bring its commercialization to a complete halt.

The issue of third-party IPRs is usually referred to as "freedom to operate" (FTO). FTO is concerned with the products or activities of a business in relation to the IPRs, particularly the patents, owned by third parties. It is an entirely separate issue to the IPRs that are owned by the business itself, and the patentability of an invention should not be confused with the freedom to use that invention.

Because it is such an important issue in determining how a technology is commercialized or indeed whether it is commercialized at all, FTO should command the attention of anyone trying to commercialize a new technology from an early stage. If a problematic third-party IPR is identified early, then it may be possible to revise or alter the planned product or activity so that it avoids the third-party IPR. For example, if a specific reagent in a chemical process is found to be covered by a third party IPR, then it may be possible to modify the chemical process to use an alternative reagent that is not covered by the third-party IPR. However, as time progresses and money and resources have been invested in finalizing and optimizing a specific product or process, re-designs and work-arounds may become prohibitively expensive.

Third parties seldom advertise the existence of their IPRs, and the only way to identify relevant third-party IPRs is to perform a search. Unlike validity searches, FTO searches only cover granted patents that are in force and patent applications that may subsequently become granted patents. Expired patents and abandoned applications do not give rise to enforceable rights and so can be ignored for FTO purposes. Even so, conducting FTO searches worldwide is a massive undertaking, especially for non-English-speaking countries, and it may be more practical to limit the search to certain countries, such as those in which some commercial activity is envisaged or where known competitors are based.

FTO searches of any scale are a specialist task that is best left to patent searchers with the relevant expertise and experience. Typically, a search will involve interrogating databases with keywords, and applicant and inventor names such as names of competitors or known experts in the field, and simply scrolling through the relevant subject-matter classification. Even with a well-designed search strategy, this process is likely to result in a large number of hits, many of which will be irrelevant. Iterative rounds of filtering are likely to be required to produce a final list of patents that need to be considered in detail [38].

At the outset of a project, a low-resolution patent landscaping exercise may be appropriate to identify the key players in the field and their specific areas of interest. This landscaping may highlight active areas where there are lots of IPRs and allow a project to be steered away from those areas from the outset. FTO searching should therefore be carried out regularly as the technology crystallizes into a defined product or process to ensure that potentially blocking rights are identified. Even if irrelevant at the outset of the development process, changes and modifications may cause IPRs to become relevant as development progresses. Furthermore, because patent applications are not published until 18 months after the initial filing, relevant patent applications may only be published and therefore identifiable in FTO searches after the initiation of the technology's development.

As with any search, it is not possible to guarantee that all potentially relevant third-party IPRs have been identified. However much FTO searching is performed there is always some possibility that your sustainable chemical technology business will run into FTO issues at some point.

If a third-party IPR is identified that might block the commercialization of a new technology, a business has various options.

One option is to modify the product or process to avoid a third-party IPR, especially if the IPR is identified early in the development process.

Another option is to reach an agreement with the owner of the third-party IPR. This will involve some negotiation and may result in a license under the IPR or the outright sale of the IPR. While either outcome involves costs, an agreement will provide a high level of certainty that the IPR is no longer a problem. Of course, it is open to the owner not to license or sell the IPR, in which case other options may need to be explored.

A final option might be to initiate legal action either at a patent office or court to invalidate a problematic third-party IPR.

4.6.2 Clearing Obstructions

Before a patent reaches grant, many patent offices allow anybody to file prior art documents that might affect the patentability of the invention. The patent examiner then considers these documents when the application is examined, along with the prior art documents identified by the patent office. Some patent offices (such as the USPTO) restrict the timing and content of submissions by third parties, but in other offices there is little restriction on what can be filed and when.

In addition, many patent offices, including the EPO and the USPTO, have procedures that allow patents to be invalidated after grant. For example, the EPO has an opposition procedure [39] that must be initiated within nine months of grant, but allows an opponent to invalidate a patent across the whole of Europe in a single proceeding. The post grant review (PGR) procedure [40] at the USPTO is broadly analogous to a European opposition procedure, and the USPTO also operates an *inter parties* review [41] procedure that is available after the nine-month period for initiating PGR.

Patents can also be invalidated through national courts at any time. Patent office proceedings are generally much cheaper than court proceedings, when they are available. However, when patent office proceedings are not available, for example because the appropriate time limits have expired, then an invalidity action through the courts may be the only option.

4.6.3 Litigation

IPRs such as patents are legal rights. This means that the courts of whichever country granted the IPR have the final decision about whether a patent in that country is valid and/or infringed.

A sustainable chemical technology business may initiate legal action to invalidate an IPR owned by a third party that blocks its activities or to enforce its own IPR against a competitor to prevent it from bringing a rival product onto the market. Alternatively, a business may find itself on the other end of legal action if a third party believes that its IPRs are infringed by the activities of the business.

In many countries, courts will often consider issues of both infringement and validity in the same proceedings, since a typical IP dispute will involve one party claiming that its IPRs have been infringed and the other party counterclaiming that the IPRs are invalid.

Invalidity issues in legal action generally focus on whether the patent meets the criteria for patentability. For example, a party may allege that the patent lacks novelty or is obvious over the prior art. Although these issues will already have been considered by the patent office in granting the patent, the party may submit new evidence to the court that was not available to the patent office and provide expert witnesses to support its position. For example, a party may base its allegations on invalidity on new prior art documents or public disclosures of an invention that were not available to the patent office. Because the evidence before the court and the manner in which it interprets the patent are often different from those in patent office proceedings, courts often find patents to be invalid, even though they have been granted by a patent office.

Infringement issues in legal action concern whether the activities of a party fall within the scope of protection of the other party's patent. The activities that can infringe a patent are broadly defined and may include making, disposing or offering to dispose of, using or importing a product, or using or offering to use a process. To infringe a patent, these activities must be carried out in the country in which the patent is in force. For example, activities performed in Canada cannot infringe a UK patent. There are generally certain activities that are legally exempt from infringement. These legal exemptions and their interpretation vary from country to country but may include private and noncommercial activities, pure research and activities relating to clinical trials, and regulatory submissions for new drugs.

If a patent is found to be infringed, courts can impose a range of sanctions on the infringing party. These may include the payment of damages to the owner of the patent and issuing an injunction to stop the infringing activity.

The stakes in any legal action are high. A party found to infringe a patent may not only have to pay significant sums in damages to the patent owner but may also be prevented from further activities that fall within the patent. This could halt the commercialization of a new technology in its tracks and may have serious consequences for the business of the infringing party. In some jurisdictions, such as England and Wales, the losing side in a court action may also be ordered to pay some or all of the winning side's legal costs. This means that trying and failing to enforce IPRs can carry significant financial penalties for the patent owner, if the other side is held not to infringe or the IPR is held invalid by the court.

Because the stakes are high when a dispute is taken to court, the parties frequently engage in negotiation both before legal action and during the legal procedure. In some cases, the threat of legal action may be enough to cause a party to change its plans to avoid a dispute. In other cases, a party may fold when it appreciates the strength of the other party's case. In practice, the majority of disputes are settled before a case reaches court.

The details of legal procedure differ from country to country and may also vary between different courts in the same country. Typically, the initial phase involves the parties setting out their case in writing. This is sometimes followed by an evidence gathering phase, followed by a trial at which the court hears the arguments and evidence on all of the issues in dispute and then decides in favor of one party or the other. If the court finds against a party at the trial, most legal systems allow the party to appeal to a second instance court (e.g. UK Court of Appeal). Sometimes, a second level of appeal is also available (e.g. the UK Supreme Court), but this is generally reserved for specific issues of law.

A number of procedures exist to help parties reach a settlement without the need to pursue a legal action to its conclusion at a trial. These procedures differ from country to country and are collectively termed "alternative dispute resolution" (ADR) and include arbitration, conciliation, and mediation [42]. Often these procedures involve a third party, such as a mediator, whose role is to either facilitate agreement between the parties or provide a decision on the issues binding on the parties. Although not suitable for every case, ADR procedures are generally cheaper and quicker than legal actions and can be more flexible in providing practical solutions to disputes that may not be available from a court. In many countries, parties to a dispute are actively encouraged to consider ADR before or during the progress of a legal action to encourage settlement.

Glossary

Applicant	A party that has initiated the process of applying for a patent or other IP right.
Claim	A series of numbered statements that precisely defines the invention in a patent application and hence defines the scope of the monopoly provided by the patent after grant. Often the subject of alteration and revision during the application process and postgrant proceedings.
Description	The section of the formal application (along with the claims and drawings) that describes the invention for which the grant of a patent is requested.
EPO (European Patent Office)	A regional patent office that grants patents on behalf of 38 European countries.

Examiner	A professional employed by a patent office to assess whether a patent application meets the criteria for patentability.
Grace period	A period before the priority date of a patent application (usually 6 months or 12 months) during which disclosures of the invention by the inventors do not count as prior art and do not invalidate the patent application. The details vary from country to country, and some countries (notably European countries) do not recognize any grace period at all.
Infringement	An act carried out without permission of the patent owner that falls within the scope of protection that is conferred by a patent that is in force.
IPR (intellectual property right)	A legal right owned by a party that covers an intangible asset. There are various different types of IPRs that protect different types of assets.
Patent	A time-limited legal right to exclude third parties from the invention defined in the claims of the patent.
Patentee	The owner of a patent.
Patent office	An organization set up to grant patents based on applications that meet the relevant criteria. Patent offices may be national (e.g. UKIPO or USPTO) or may be regional (e.g. EPO).
Patent application	A pending request for the grant of a patent covering the invention that is described in the application document. A patent application is not an enforceable legal right until it is granted (and many applications are never granted), but in some circumstances rights acquired after grant can be backdated to the date of publication of the patent application.
Patent agent	A professional who is qualified to prepare, file, and prosecute patent applications on behalf of clients. The term is synonymous with "patent attorney" in some jurisdictions, such as the UK. In the US, a patent agent is a practitioner who has passed the patent bar examination and is registered to practice with the USPTO.
Patent attorney	A professional who is qualified to prepare, file, and prosecute patent applications on behalf of clients. The term is synonymous with "patent agent" in some jurisdictions, such as the UK. In the US, a patent attorney is a qualified legal practitioner who can practice before the courts in at least one US state, as well as at the USPTO, and is able to advise on a range of IP matters.

PCT (Patent Cooperation Treaty)	An agreement that allows a single patent application (a "PCT application") to be filed that can subsequently be nationalized in any one of 152 contracting states. This has the significant advantage of delaying the costs of this nationalization by about 18 months and also provides for a search and examination opinion that gives an idea of the prospects of a patent being granted on the application.
Prior art	All the information that has been made available to the public before a certain date (generally the priority date), against which the patentability of an invention is assessed.
Priority	A system governed by an international agreement (the Paris Convention) that allows a person who has filed a first patent application for an invention (a "priority filing") to file a second patent application for the same invention within 12 months of the first application that is entitled to retain the date of filing (the "priority date") of the first patent application. This means that the patentability of the second application is assessed only with respect to information known to the public before the date of filing of the first patent application.
Publication	A step in the patent application process in which the formal application document is published. This generally occurs 18 months after the earliest priority date that is claimed by the application.
Search	A step in the patent application process in which all the information that was available to the public before the priority date of a patent application is reviewed to try to find publications or other material that might invalidate the patent application.
USPTO (US Patent and Trademark Office)	The patent office that grants patents for the US.
Validity	The issue of whether or not a patent meets the criteria for patentability and has therefore been validly granted. The individual criteria may be reviewed in postgrant proceedings at a patent office or in court proceedings to assess the validity of a patent.

References

1 Bridge, M. (2015). *Personal Property Law*. Oxford: Oxford University Press.
2 Bently et al. (2018). *Intellectual Property Law*, 5e. Oxford Univ Press.
3 Vestergaard Frandsen A/S & Ors v Bestnet Europe Ltd & Ors, UKSC 31 (2013)

4 European Union Directive 2016/43

5 *Waymo LLC v. Uber Technologies Inc.*, No. 17-2235 (Fed. Cir.) (2017)

6 Art 52(1) European Patent Convention

7 35 U.S.C. § 101 (1952)

8 Gov.UK. (n.d.). Green channel. https://www.gov.uk/guidance/patents-accelerated-processing#green-channel (accessed 9 November 2020).

9 Canadian Intellectual Office. (2020). Cleantech and intellectual property. https://www.ic.gc.ca/eic/site/cipointernet-internetopic.nsf/eng/wr04431.html (accessed 9 November 2020).

10 Paris Convention for the protection of industrial property (1883)

11 Patent Cooperation Treaty (1970)

12 European Union Regulation (EC) No. 469 (2009)

13 Drug Price Competition and Patent Term Restoration Act of 1984, Public Law 98-417, 98 Stat. 1585 (codified at 21 U.S.C. 355(b)

14 Manual of Patent Examining Procedure 2733

15 Guidelines for Examination in the EPO G II 3

16 Manual of Patent Office Practice 12.03

17 Ariosa Diagnostics, Inc. v. Sequenom, Inc., 788 F.3d 1371 (Fed. Cir. 2015)

18 Alice Corp. v. CLS Bank International, 573 U.S. 208 (2014)

19 Guidelines for Examination in the EPO G VI

20 Manual of Patent Examining Procedure 2131 (2018)

21 Manual of Patent Examining Procedure 2153 (2018)

22 Manual of Patent Office Practice 15.04

23 Guidelines for Examination in the EPO G VII

24 Manual of Patent Examining Procedure 2141 (2018)

25 Guidelines for Examination in the EPO F III

26 Manual of Patent Examining Procedure 2164 (2018)

27 EPO Enlarged Board of Appeal Decision G2/98

28 Guidelines for Examination in the EPO F VI

29 Art 4 Paris Convention (1883)

30 Patent Cooperation Treaty Applicant's Guide (WIPO)

31 Art 15 Patent Cooperation Treaty

32 Art 18 Patent Cooperation Treaty

33 Manual of Patent Examining Procedure 2500 (2018)

34 Hunt, Nguyen, and Rodgers (eds.) (2007). *Patent Searching Tools and Techniques*. Wiley.

35 Trade Secrets: Law and Practice Quinto, Singer and McCauley LexisNexis 2014

36 Guidelines for Examination in the EPO E VIII 4

37 Manual of Patent Examining Procedure 708 (2018)

38 Patent Freedom to Operate Searches, Opinions, Techniques, and Studies, A Zuege American Bar Association (December 7, 2019)

39 Guidelines for Examination in the EPO D

40 US Patent and Trademark Office. (2020). Post grant review. https://www.uspto.gov/patents-application-process/appealing-patent-decisions/trials/post-grant-review (accessed 9 November 2020).

41 US Patent and Trademark Office. (2020). Inter partes review. https://www.uspto .gov/patents-application-process/appealing-patent-decisions/trials/inter-partes-review (accessed 9 November 2020).

42 Blake, S., Browne, J., and Sime, S. (eds.) (2018). *A Practical Approach to Alternative Dispute Resolution*, 5e. OUP Oxford.

Part II

Political and Environmental Considerations

5

Navigating and Leveraging Government Entrepreneurial Ecosystems for Support

Janine Elliott[1] and Rohit Sood[2]

[1] Los Angeles Cleantech Incubator, Los Angeles, CA, USA
[2] Spinverse Oy, Espoo, Finland

5.1   What Is an Entrepreneurial Ecosystem?

All over the world, local and national governments are working with nonprofit and for-profit organizations to build *entrepreneurial ecosystems*. Like a natural ecosystem, entrepreneurial ecosystems share similar traits.

- Cycling of resources through the system. Natural ecosystems cycle water and nutrients; entrepreneurial ecosystems cycle money and expertise.
- Competition for resources between groups. Just as seedlings reach for limited sunlight, start-ups enter pitch competitions for limited prizes.
- Specialized niches efficiently use resources. In nature, many insects eat specific types of plants only; in entrepreneurial ecosystems, different organizations and groups are specialized to work with certain types of start-ups.
- Symbiotic relationships between groups. In a natural ecosystem, lichen is formed by a mutually beneficial connection between algae and fungi. In an entrepreneurial ecosystem, commercialization programs might be co-created and run by a nonprofit organization, a for-profit incubator, and corporate sponsors that have a common interest but different areas of expertise.

This chapter will help innovators of sustainable chemical technologies identify what resources they may need, who the players are in the ecosystem, and what resources those entities have to offer. It is structured into four sections.

- Types of resources available
- Ecosystems in the US and Canada
- Ecosystems in the EU
- Setting priorities when pursuing resources

Although many other regions and countries have their own ecosystems in place, these will not be covered in this chapter. However, the basic concepts still apply, and readers are encouraged to use these as a foundation for finding resources appropriate for them.

How to Commercialize Chemical Technologies for a Sustainable Future, First Edition.
Edited by Timothy J. Clark and Andrew S. Pasternak.
© 2021 John Wiley & Sons Ltd. Published 2021 by John Wiley & Sons Ltd.

5.2 Types of Resources Available

First, there are several resources an innovator may require, such as funding, space, training, services, and networks. Next, there are different types of organizations that offer these resources, such as government agencies, for-profit companies, and nonprofit organizations. Finally, there are logistical considerations that should help innovators prioritize: discerning eligibility based on mandates of the entities (such as the stage of the enterprise's development, geographic location, or type of environmental impact that will be achieved) and deciding whether the timing or trade-offs of pursuing them are appropriate.

5.2.1 Financial Resources

One resource for the innovator usually rises above others in terms of priority: money! Financial support for sustainable chemical innovations is often a mix of different types of capital, several of which are described in the following sections.

5.2.1.1 Grants

Proposals are submitted to a funder and, if awarded, the innovator does not need to pay the money back. However, they will be bound by certain reporting requirements. Some grant applications are offered at regular intervals, some are a special request for proposals (RFPs) on a particular topic, and some are not publicized and are requested by invitation only.

5.2.1.2 Prizes

Innovators compete in pitch or business plan competitions and can win prize money. It does not need to be repaid and tends to be the most flexible funding source. However, the prize amounts can be relatively small compared to the effort. Because of this, be sure there are networking or learning opportunities that would help your venture even if you don't win.

5.2.1.3 Investments

Innovators with capital-intensive technical development might start with grants and prizes, but they eventually need large amounts of money to scale their technology and business. The risk of losing money is high for early-stage investors, so investors purchase a portion of ownership, known as "equity," in the company with hopes that it becomes valuable in the future. Venture capital (VC) firms, angel investors, and corporate venture groups are well-known investor types. However, philanthropic, nonprofit, and government groups also invest in sustainable chemical technologies; details about how these emerging investors plug into the ecosystem are in the next section. Chapter 12 contains information on financing sustainable chemistry start-ups.

5.2.1.4 Loans

In most cases, start-ups will not be eligible for a loan from a bank or government program as they are not yet profitable and likely do not even have established revenue streams. These types of funders seek companies with lower-risk profiles. For these more established companies, a national government program may offer low-interest loans or loan guarantees for infrastructure to scale operations.

5.2.2 Nonmonetary Resources

In addition to money, there are many types of resources that help successful ventures advance from the lab to the market.

5.2.2.1 Training

Training programs, bootcamps, workshops, and accelerators are invaluable to anyone who wants to learn how to get a product or process to market. Sustainable chemistry innovators are often brilliant scientists, but typically they need help building their knowledge of business and entrepreneurship. Programs can last anywhere from one day to six months or more and may be in-person, virtual, or a mix. To get started, local introductory workshops are helpful in building basic business acumen and are usually broad enough to appeal to many types of ventures https://www.greentownlabs.com/about/wetlab. 3 Day Startup (www.3daystartup.org) is a great example of programming at many college campuses. Then, as you become more sophisticated, look for workshops or accelerators that specifically recruit for deep tech, hard tech, or industry-specific innovators because they will have a more nuanced understanding of how to help you progress toward commercialization. For example, VentureWell's ASPIRE training is ideal for deep tech ventures (https://venturewell.org/aspire), and Village Capital's Ag & Food program (https://vilcap.com/sector/agriculture) is great for sustainability-oriented startups interested in agricultural applications of their technology.

5.2.2.2 Space

Incubators, accelerators, and co-working spaces are increasingly common in major metropolitan areas and small cities too. They offer a place to convene a team with like-minded entrepreneurs and to hold meetings with customers or investors in a professional atmosphere. Though most of them only offer office space, some are beginning to offer prototyping, lab facilities, and pilot-plant space as well (https://greentownlabs.com/greentown-boston/wet-lab/). In addition to physical space, incubators that are best-suited to sustainable chemistry-based ventures will offer events and access to technical advisors, industry-savvy investors, and third-party testing capabilities that can increase credibility and expedite time to market.

5.2.2.3 Advice, Coaching, and Professional Services

Entrepreneurs can benefit greatly from "in-kind" or free services offered through an accelerator, incubator, or community events. Advisors or coaches volunteer their time to a particular program and will be assigned to you to help address challenges specific to your business. In other cases, you may be introduced to experts who donate some of their professional services that start-ups typically cannot afford. This might include advice on building your team, setting up a corporation, accounting, managing intellectual property, and more. Accelerators, incubators, nonprofit organizations, and government offices may offer connections to these resources who are driven by a desire for you to eventually become a paying client. For example, the Los Angeles Cleantech Incubator (LACI) matches member companies with executives in residence (EIRs) that offer coaching and accountability based on startups' strategic needs, in addition to training, networking, and work-space resources.

5.2.2.4 Networks

Ever heard the expression "It's not what you know, but who you know" or "You don't know what you don't know"? Opportunities to build professional networks are critical for early-stage innovators because most of the commercialization process will actually happen *outside* the lab. For many technically oriented innovators, networking events can be awkward or intimidating at first. But, if you keep attending them to build your skills and refine your pitch, it will get easier to talk with people you don't know.

Networking events are a great way to learn about your industry and customers, refine your business ideas and pitch, meet potential teammates and advisors, and discover more resources of which you weren't aware. These are easily found by browsing Eventbrite.com or Meetup.com.

5.3 Ecosystems in the United States and Canada

The next dimension of navigating the ecosystem is to recognize which type of entity is offering the resources and why. In the US/Canada, these groups include government agencies, nonprofit organizations, for-profit service providers, investors, academic research institutions, and more. They will often work together to support a pipeline of innovators who travel on the trajectory from proof-of-concept in the lab all the way to commercialization. A summary chart is provided at the end of the chapter.

Even though these entities may offer similar resources, their motivations for participating in the ecosystem and their ideal entrepreneurial clients may differ greatly. If you understand the players and their motivations, you can make strategic decisions about what resources to pursue, gauge your likelihood of getting what you want, and evaluate the trade-offs.

5.3.1 Government Agencies

Government agencies at the national, regional, and local levels are funded by tax revenue. The overall goal of all of these agencies is to support entrepreneurs who will create jobs and generate enough revenue to eventually become tax-paying businesses. Government agencies may also have a specific mandate to meet environmental or social policy goals. They are likely to offer funding programs directly to start-ups at many different stages or sponsor start-up competitions in partnership with other ecosystem players to support innovation. Though they want to steward tax dollars carefully, they are making longer-term, indirect investments in the future.

5.3.1.1 State-Level Resources in the US

Search your state's .gov website to find economic development resources for start-ups that are located within that state. Many states offer start-up-oriented resources that support other policy initiatives, such as meeting clean energy goals or supporting the local agricultural economy. This can make your search a little easier, but it is good to seek out opportunities that aren't specifically "green" as well.

For example, one of the most comprehensive state-level groups supporting innovations in sustainability is the Massachusetts Clean Energy Center (MassCEC). It is a state economic development agency dedicated to accelerating the growth of the clean energy sector to spur job creation, deliver state-wide environmental benefits, and secure long-term economic growth for their citizens (https://www.masscec.com/about-masscec). They provide a variety of grants and investments directly to entrepreneurs and researchers while also supporting local incubators and accelerators through grants and event sponsorships. Therefore, a Massachusetts-based start-up applying for MassCEC funding likely wants to demonstrate its fit with their mandate by articulating how their sustainable chemical technology will save water or energy compared to incumbent technologies and how the eventual growth of the business will lead to job creation in the state.

Other leading state agencies that have a track record of supporting sustainable chemistry companies include the California Energy Commission's CalSEED program and the New York State Energy Research and Development Authority.

5.3.1.2 National Resources in the US

The US federal government offers a plethora of government sponsored research, business plan competitions, training, and other resources to spur scientific innovation. The following is a small sample of federal initiatives that are likely to be relevant to deep tech innovators in the US:

- The Small Business Administration is an agency that coordinates the Small Business Innovation and Research and Small Business Technology Transfer Research grant programs (SBIR/STTR) across 15 federal agencies. These include the US Department of Agriculture (USDA), National Science Foundation (NSF), Department of Energy (DOE) and others, with grants ranging from $100 000–750 000 and beyond. The SBIR/STTR grants help innovators conduct activities that will increase the commercialization of their technologies. However, each agency has a slightly different grant application process and different goals.

  For example, the NSF SBIR program focuses on research and development for a wide variety of technologies that have a clear commercial application but are still too risky for private investors. As the agency's motivation is rooted in job creation and increasing the tax base, you may increase your chances of being funded by detailing how the grant will help you reach a technical milestone within 6–24 months that will trigger investors and customers to fund your next stages of development toward scale-up. Note that the novelty of your sustainable chemical innovation is only one important facet of the grant application.

  Other federal agencies may have a different approach to SBIR/STTR proposals, so it will be important to explore your options. For example, the USDA's SBIR (https://nifa.usda .gov/program/small-business-innovation-research-program-sbir) grants are understandably focused on agriculture-specific priorities, such as technologies that valorize agricultural waste or reduce post-harvest food waste. It is likely that successful applicants of these government grants will have clearly described both how their business could eventually succeed as a US tax-paying employer and how deployment of the technology could impact American agricultural production, nutrition, and/or rural incomes. In addition to other eligibility and evaluation criteria, be sure to carefully review any agency's most recent request for applications solicitation (https://nifa.usda.gov/sites/ default/files/rfa/fy-20-21-sbir-program-phase-i-rfa-20190725.pdf). They often welcome sustainable chemistry technologies in a variety of categories, but it is unlikely to be its own category – think creatively when reading the topic areas.

- The NSF also sponsors the Innovation Corps (I-Corps) program, which offers grant funding and rigorous training programs for innovations emerging from academic and federally funded research labs. It is ideal for teams who are in the earliest stages of commercialization and are identifying potential customers and partners (https://www .nsf.gov/news/special_reports/i-corps/webinars.jsp). The training develops important customer engagement skills that helps teams understand the core values and necessary technical requirements of customers. For example, free online resources include: https:// venturewell.org/i-corps/llpvideos and https://www.udacity.com/course/how-to-build-a-startup--ep245. As a result, I-Corps graduates who find a product–market fit tend to prepare successful funding proposals, such as SBIR applications and business plan competition presentations [1].

5.3.1.3 National Resources in Canada

There are a number of national-level funding organizations in Canada that the sustainable chemistry innovator can potentially access. These include the following:

- The National Research Council (NRC) Industrial Research Assistance Program (IRAP) provides funding to many different types of small and medium-sized companies in the early stages of commercialization, generally before scale-up. NRC IRAP offers financial resources for a start-up's technical development and access to specialized business and technical services. In addition, as a government agency dedicated to wealth creation and small Canadian businesses, they also help start-ups plan for international collaborations and export (https://nrc.canada.ca/en/support-technology-innovation), which can be an important component of the long-term business model for many chemical innovations.
- Sustainable Development Technology Canada (https://www.sdtc.ca/en) provides funding for innovators who clearly articulate how their technology addresses climate change, clean air, water, and soil so that Canadian innovators can be world leaders while also benefitting the national economy. SDTC funds companies scaling to the pilot or demonstration phases and require an industrial strategic partner to provide a portion of the funds.
- Strategic Innovation Fund (SIF) provides large-scale (>$10 million) funding to companies across the entire technical development spectrum in many industries (https://www .ic.gc.ca/eic/site/125.nsf/eng/home). Their primary mandate is to increase the rate of innovation and technology transfer in Canada, help Canadian companies grow globally, attract private investment, and support the creation of networks of academic, nonprofit, and for-profit actors. Therefore, chemical innovators should note funding calls that usually focus on specific industries or applications, in addition to exploring opportunities among their five funding streams.

These examples reflect a small sample of the agencies and nationally funded programs that exist to support start-ups in the US and Canada. To start looking for resources relevant to you at the country level in the US, the US Small Business Administration will lead to several different funding opportunities. Canadian start-ups should consider starting their search with Canada's Regional Development Agencies (RDA) to locate nationally funded regional opportunities.

5.3.2 Non-profit Organizations

These organizations support entrepreneurs who impact a particular environmental or social issue and are less concerned with financial payback than a typical business or investor. Their programs may prioritize support for entrepreneurial women, technologies that increase environmental sustainability, businesses that address climate change, and so on; their strategy might have other stipulations such as growth stage or geographic location as well. Nonprofit organizations are usually funded by individuals, government agencies, and/or philanthropic foundations who have an aligned interest in the nonprofit's mission. Nonprofits use that funding to offer programming, training, services, and direct funding to innovators. The more you can articulate how you fit into the work that they do (and, implicitly, what their funders are looking for), the more likely it is that they will support you over other innovators.

5.3.2.1 Sustainable Chemistry-Specific Non-profits

Sustainable chemistry-specific nonprofits are somewhat rare in the US and Canada at the time of writing, but the following are a few pioneers:

- GreenCentre Canada (https://greencentrecanada.com) offers early-stage sustainable chemistry companies access to a highly trained team of chemists and chemical engineers who provide technology development services in fully equipped lab facilities. All intellectual property is assigned to the client company, and these services, including access to numerous networks, are offered at no cost to the entrepreneur within its public programs. Similar services are also offered on a fee-for-service basis to larger, more established companies.
- Think Beyond Plastic "harnesses the forces of innovation and entrepreneurship for a world free of plastic pollution" – from food packaging to textiles and ocean pollution (https://www.thinkbeyondplastic.com). Its Innovation Center offers space, networks, and professional guidance, while its Annual Innovation Challenge offers an accelerator program that specifically recognizes the importance of sustainable chemistry and the development of bio-benign materials.
- The Green Chemistry and Commerce Council (GC3) fosters collaborations and promotes best practices for green chemistry innovation across multiple groups – large companies, government agencies, nonprofits, start-ups, and academic researchers. Its membership-based Startup Network offers educational webinars, networking events, and opportunities to showcase market-ready green chemistry technologies.

5.3.2.2 Other Sustainability and Social Impact Organizations

These organizations do not *exclusively* serve sustainable chemistry innovators but are still interested in working with start-ups provided you articulate how your innovation fits with their particular mission or program. There are lots of them, including the following:

- VentureWell is a nonprofit that offers grants and training programs for innovators in science and technology (venturewell.org). Its E-Team program gives $5000–$20 000 grants to entrepreneurial students in the US who are developing science-based innovations that can have a large impact on people and the planet. Here, a successful proposal goes beyond detailing why the technology is compelling to potential customers to also include how it will have significant environmental benefits or improve human health. For example, past grantee Hazel Technologies was created by students at Northwestern University, who proposed that its nontoxic chemical sachets may slow the metabolism of fruit so they do not rot in shipping. This would allow farmers and grocers to benefit financially while society benefits from reduced greenhouse gas emissions that would have been caused by the food waste. You can read more about them as a "success story" in Chapter 14.
- Echoing Green is a nonprofit that takes a different approach by focusing not on the technology or industry, but on how an idea can address a major societal issue. Ambercycle is an example of an early-stage company that successfully connected its sustainable chemistry innovation to new ways to address the problems of ocean pollution and plastic recycling. As a result, they received funding, personalized leadership-building skills, and access to a network of funders and partners. This example demonstrates that even though an opportunity might not immediately appear to be a fit – after all, it's a big jump from the lab bench to saving the oceans! – sometimes it pays off to think big when seeking out resources.

5.3.3 Incubators and Accelerators

A variety of incubation and acceleration resources are offered by for-profit companies and nonprofit organizations. These often offer a mix of resources for start-ups, such as mentorship or access to networks. Though the terms are often used interchangeably, this section clarifies a few distinctions.

Incubators' most dominant trait is that they offer workspace and opportunities to network in exchange for rent, a membership fee, or a portion of ownership in the start-up. They tend to fall into two categories: private for-profit company or public-institution affiliated nonprofit. They do not differ dramatically, but it may impact the types of services offered and level to which fees may be subsidized. In addition to desk space and meeting rooms, incubators increasingly offer maker-spaces for physical prototyping, and a few are now offering wet lab space. Greentown Labs, near Boston in the US, is a field-leading example of a privately owned incubator that offers office, lab, and maker-space while also providing a series of educational and networking events for its cleantech-focused community of members and advisors.

Accelerators' most dominant trait is that they offer investment funding in exchange for equity in the innovators' company, and to enhance their investment, they offer an accelerator program that equips entrepreneurs with skills to build their companies. An alternative interpretation is that accelerators offer venture development training and financial resources so that teams can dedicate their full attention to the venture without having to worry about personal expenses. One of the most famous is TechStars. It is an accelerator that operates in more than 150 countries worldwide, offers a three-month in-person program with a seed investment, emphasizes mentorship over a set curriculum, and provides connections to investors and alumni in a robust network. Though it does not work with sustainable chemistry start-ups exclusively, Elegus, a battery component developer, and Mobius, a bioplastics company, have successfully won places in TechStars programs.

5.3.4 Academic Research Institutions

Universities have always been a source of fundamental research and groundbreaking inventions. They are also playing an increasing role in assisting innovators translate their technologies into products or processes ready for commercialization. These institutions may offer a mix of incubation, training, fabrication space, and third-party testing to ventures at a variety of stages of technical development.

As you review the following examples, be sure to explore similar resources that may be offered by institutions near you:

- The University of Massachusetts at Lowell (UML) has built an ecosystem around its campus (https://www.uml.edu/Innovation-Hub/What-We-Offer) so that different types of ventures can take advantage of various resources based on their needs.
  - The Innovation Hub is an incubator space with two locations and several tiers of membership and prototyping center privileges. It is available to both UML community members and the general public.
  - The Prototyping and Fabrication Lab enables start-ups to validate and scale their ideas. For a chemical innovator, this may be a space to model designs for larger-scale reactors as you move beyond the bench-scale testing phase.
  - UML's Core Research Facilities (https://www.uml.edu/Research/CRF), which includes fee-for-service and licensing arrangements, are ideal for companies that want to take advantage of equipment and space without buying their own. Examples include a clean-room nanofabrication lab, high-tech textile development at the Fabric Discovery Center, and the Materials and Biomolecular Characterization Labs.
- Velocity (https://velocityincubator.com/about-velocity) at the University of Waterloo in Canada offers a comprehensive set of resources for sustainable chemistry start-ups. As both an academic institution and government-funded hub for economic development, it accepts start-ups from around the world, not just those spun out of the university or even based in Canada. In addition to networking opportunities and office space typical of most incubators, it offers support for technical development, such as chemical/biological labs and piloting space, capital equipment, prototyping, and hardware support.

Universities may also offer competitions that award prizes, "acceleration" training, mentorship, and/or publicity to start-ups emerging from the community. For example, at UML, the Difference Maker competition and training is for UML students working to solve real-world problems, and the Idea Challenge awards prizes up to $50 000. Velocity's Student Grant Pitch competition offers investment up to $50 000 too.

Note that it will take some searching to find the right academic resource for you. Services may be offered at a central location that acts as a "clearinghouse" for resources (as with Velocity) or within various colleges or centers across campus.

Finally, consider the financial incentives of academic-affiliated groups and whether the trade-offs are appropriate for you. Depending on the resource offered and your existing affiliation with the institution (e.g. if you are a registered student), there may be a fee to access resources. If you are eligible, a capital-efficient way to hit key technical milestones is to use grant or prize money. However, many universities that provide technical support will claim ownership of any new intellectual property – *even if you are paying their bills!* This can introduce major complications for engaging future investors if not carefully negotiated with the university. Make sure the intellectual property policy is fully understood and will work for your company's strategy.

5.3.5 Investors

Investors are an important part of the entrepreneurial ecosystem because they provide capital that allows early-stage start-ups to grow. Financing options are described in detail in Chapter 12.

Investors can also be amazing networking and mentorship resources. As you attend networking events and training sessions, you will meet investors. They will often volunteer their time as a mentor, panelist, or competition judge because they are scouting for new deals. In a networking or mentoring context, don't pitch to them assuming you'll magically attract their investment. Instead, find out whether you're a good match. Ask them about the types of companies they invest in, what makes a compelling deal, how big the deals usually are, their technical background, and so on. Not every investor is looking for a deal with a company like yours, and that's OK! They might know someone else who can help you. Instead, they may have advice about building a company in your industry. Even better, they might be interested in a company like yours but want to see you hit certain technical milestones or build up your team first (https://www.businessinsider.com/invest-in-lines-not-dots-2010-11). Strengthen your network with these contacts and send them occasional updates so you can approach them later when the timing is right.

You may be ready to raise investment once you have taken advantage of many of the previously mentioned resources to build your team while reducing technical and commercial risks. You are more likely to have success raising investment from VCs or angel groups if you understand their *thesis*: most investors will have a specific area of expertise and will focus on certain industries, technologies, or geographies. For example, Rhapsody Venture Partners and Pangaea Ventures specialize in materials technologies, while Chemical Angels are interested in different chemical technologies, including sustainable chemistry. These groups often bring insight into scale-up and go-to-market plans that would elude less-specialized investors. At the time of writing, SaferMade is the only VC firm in the US that is solely focused on green chemistry.

5.3.6 Hybrids of Resources and Players

There can be significant overlap and nuance in the entrepreneurial ecosystem between government agencies, nonprofit organizations, for-profit accelerators, incubators, academic institutions, and investors. A few examples include the following:

- Nonprofit organizations joining the investment scene with novel funding models that give priority to environmentally positive start-ups, while tolerating the high risks inherent in early-stage ventures. Example include the following:
  - ○ VentureWell is a nonprofit that makes investments in certain graduate companies of its ASPIRE accelerator, which supports science-based innovators making a positive social or environmental impact.
  - ○ The PRIME Coalition is a nonprofit that partners with philanthropists to invest in companies that combat climate change, have a high likelihood of achieving commercial success, and would otherwise have a difficult time raising sufficient financial support.
  - ○ Government agencies – like MassCEC in the US and Innovation, Science, and Economic Development Canada that administers the SIF program – are using a similar investment models. With expectations somewhere between nonprofits and private investors, the public funds incentivize private investors to join a syndicated round of financing so that they feel more comfortable spreading the commercial risk.
- Oak Ridge National Laboratory's Innovation Crossroads and Cyclotron Road at the Lawrence Berkeley National Laboratory offers a fellowship program funded by the US Department of Energy, other government agencies, academic institutions, and several corporate sponsors. More than a grant, training accelerator, or rent-charging incubator, the program offers funding, lab space, technical expertise, and other venture development services.

5.4   Ecosystems in the European Union

The EU offers broad entrepreneurial ecosystems with a variety of resources available to the sustainable chemistry innovator at two levels: (i) EU-wide and (ii) individual countries. Because of the vast number of individual countries in Europe and their respective programs, this section will focus solely on EU-wide initiatives. In any case, make sure you check what programs are available in your home nation as these could be highly beneficial.

Start-ups have been at the center of EU public policy for the past 10 years with widespread interest in improving the conditions and framework of entrepreneurial ecosystems to facilitate start-up creation and growth [2–4]. The European Commission (EC) launched the Startup Europe (SE) initiative in 2014 with the objective to make the European ecosystem more coherent [5].

By 2050 the EU wants to become the world's first climate-neutral continent. To achieve this, the EC presented the *European Green Deal* with the goal of enabling European citizens and businesses to benefit from the transition to sustainable technologies. The key objectives of the European Green Deal include the following:

- Becoming climate-neutral by 2050
- Reducing pollution and protecting human life, animals, and plants
- Helping companies to become world leaders in clean products and technologies
- Helping to ensure a just and inclusive transition

An initial roadmap of key policies to preserve Europe's natural environment ranges from ambitiously cutting emissions to investing in cutting-edge sustainable research and innovation. The European Green Deal targets improving the well-being and health of its citizens and future generations and is intended to be a new growth strategy.

5.4.1 SusChem: A European Technology Platform of Sustainable Chemistry

SusChem is a European technology forum of industry, academia, and policy makers that focus on sustainable chemistry. It was founded by six European bodies representing the main stakeholders from academia and industry in the chemical sciences sector. These include the following:

- Royal Society of Chemistry (UK)
- European Chemical Industry Council
- German Chemical Society
- European Federation of Biotechnology Section on Applied Biocatalysis
- German Society for Chemical Engineering and Biotechnology
- European Association for Bioindustries

SusChem works to create innovation priorities that can be implemented at both Pan-European and national levels. Although they do not provide direct funding, the organization takes a lead in being the voice for advancing sustainable chemistry technologies.

5.4.2 Entrepreneurial Ecosystem Resources

Small and medium-sized enterprises (SMEs), representing about 4% of the EU's GDP, and entrepreneurship are considered key to ensuring economic growth, innovation, and job creation by the EC. EU funding is available for entrepreneurs, start-ups, micro-companies, and SMEs through a range of initiatives and can be divided into (i) direct and (ii) indirect funding.

5.4.2.1 Direct Funding

Direct funding in the form of grants and contracts are managed by the EC, and entrepreneurs can apply for this funding on its Funding and Tenders web portal (https://ec.europa.eu/info/funding-tenders/opportunities/portal/screen/home). Grants are offered to start-ups and SMEs, following a call for proposals, for specific projects directly related to the interests of the EU and their programs or policies (https://europa.eu/youreurope/business/finance-funding/getting-funding/eu-funding-programmes/index_en.htm). Grants are a form of complementary financing. The EU does not usually finance projects up to 100% (meaning the start-up or another participating entity must co-finance).

Horizon 2020 with a budget of $93.4 billion (€80 billion) was the biggest EU research and innovation program available from 2014 to 2020 (https://ec.europa.eu/programmes/horizon2020/en/what-horizon-2020). This funding has advanced breakthrough innovations from the lab to the marketplace and targeted, among others, European start-ups or SMEs. Horizon 2020 grant calls for submission were aligned with the EU's social, economic, or infrastructural goals. The EU is promoting sustainability and offers grants to companies that are developing new sustainable chemical technologies.

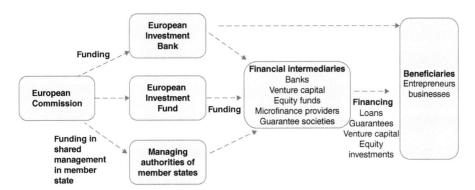

Figure 5.1 European Commission–backed financing available for entrepreneurs and businesses.

An example of a sustainable chemistry start-up that leveraged this program is Elastopoli Oy. Their technology is centered on nanocellulose composites (AquaComp) with superior properties relative to nonrenewable polymer composites currently used in automotives. During this funded project, Elastopoli achieved market replication of the AquaComp composite into a variety of industrial goods including those in the automotive market.

Horizon Europe (2021–2027) is an ambitious $116.7 billion (€100 billion) program to succeed Horizon 2020.

5.4.2.2 Indirect Funding

A wide range of indirect EU funding is available via business loans, microfinancing, guarantees, and venture capital. Every year the EU supports more than 200 000 businesses by indirect funding (Figure 5.1). Local financial institutions such as banks, venture capitalists, or angel investors decide how the funds from the EU are disbursed. These financial institutions also determine the exact financing conditions such as the amount, duration, interest rates, and fees. Entrepreneurs in the EU can contact one of more than 1000 financial institutions to find out more information.

For sustainable chemistry start-ups, these two programs are of particular relevance:

- Competitiveness of Enterprises and SMEs (COSME)
- Innovfin

5.4.3 Competitiveness of Enterprises and SMEs (COSME)

The COSME program has been running in its current form since 2014, with a current budget of $2.7 billion (€2.3 billion). It supports entrepreneurs, SMEs, and business support organizations by promoting entrepreneurship and improving both the business environment for SMEs and the international competitiveness of European enterprises.

COSME achieves this by doing the following:

- Facilitating access to finance through dedicated financial instruments in all phases of their life cycle – creation, expansion, or business transfer. This gives early-stage companies ready access to guarantees, loans, and equity capital.

- Supporting access to the broader marketplace via the Enterprise Europe Network: a "one-stop shop" for the business needs of SMEs in the EU and beyond. To increase business competitiveness, it provides information and customized business support services for free.
- Reducing the administrative and regulatory burden on SMEs and simplifying the policy environment.
- Facilitating access to markets and improving the business environment between the EU and its main trading partners by providing cooperation and reducing regulatory differences.
- Financing from loan guarantees – COSME offers business loans of up to $175 163 (€150 000) for more than 200 000 SMEs in Europe per year.

The process of obtaining financial assistance from COSME begins by searching for institutions that partner with the EC and then applying for EU financing directly through those institutions.

5.4.4   InnovFin – Financing for Innovators

InnovFin, alongside Horizon 2020, has a mission to support research and innovation throughout the EU. Most of the financing in this program is handled by the European Investment Bank (EIB) and comes in several forms such as project loans, intermediated loans, venture capital, venture debt, microfinance, and equity and fund investment. The InnovFin program is divided into two types: InnovFin Equity and InnovFin SME Guarantee.

InnovFin Equity, managed by the European Investment Fund (EIF), is an initiative launched by the EC and the EIB as part of the Horizon 2020 framework. InnovFin Equity focuses on early-stage financing of innovative enterprises located or active in the EU by providing equity investments and co-investments alongside funds allocated by Horizon 2020. Innovative sectors covered by the program include sustainable technologies, as well as addressing societal challenges including resource efficiency, the bio-economy, and climate change. InnovFin Equity is further divided into four segments: InnovFin Technology Transfer, InnovFin Business Angels, InnovFin Venture Capital, and InnovFin Fund-of-Funds.

InnovFin SME Guarantee provides guarantees worth $29 000–$8.7 million (€25 000–€7.5 million) to start-ups and SMEs. This is done through financial intermediaries, which are guaranteed or counter-guaranteed against a portion of their potential losses by the EIF. In general, eligibility for InnovFin programs is limited to innovative companies and investment firms.

5.4.5   European Innovation Council (EIC) Accelerator

The EIC Accelerator is an extension of the InnovFin program that aims to help "high-flying SMEs" develop "radical market-creating innovation to improve productivity and international competitiveness." Specifically, it rewards start-ups and SMEs that have marketable innovations ready to scale. It is deployed in three phases and is highly competitive with selected applicants decreasing substantially with each phase.

Phase 1 (concept and feasibility assessment). This phase includes up to three days of coaching and mentoring as well as up to $58 000 (€50 000) for research and development expenses. Impact, excellence, quality, and efficiency of implementation are the four main factors used to assess applications.

Phase 2 (innovation project). This phase focuses on bringing the concept from phase 1 to market. In addition to up to 12 days of further coaching and mentoring, phase 2 includes production financing (typically from $583 000–$2.9 million [€500 000–€2.5 million]). Higher levels of funding can be requested as long as applicants are able to justify them. Phase 2 applications are carried out in two stages, remote assessment based on qualitative criteria followed by panel interviews with experts from the technology, finance, and business sectors.

Phase 3 (commercialization). This phase consists of sustained investment readiness support and ongoing help to access risk finance and customers to further scale the business. By the end of phase 3, the company should have placed one or more innovations (product, process, service, etc.) on the market, boosting its competitiveness and positioning it for growth.

5.4.6 Other EU Programs for the Entrepreneur

5.4.6.1 The EC Startup Fund

With a budget of $3.15 billion (€2.7 billion) for 2018–2020, the EC funds innovators, entrepreneurs, and small companies with bright ideas and the ambition to scale internationally, through a variety of non-equity-based grants. Start-up funding will continue to be available in Horizon Europe (2021–2027), previously described.

5.4.6.2 Fast Track to Innovation (FTI)

The EIC FTI is a full bottom-up innovation support program open to industry-driven consortia that can be composed of all types of EU-based participants including entrepreneurs. It helps partners co-create and test breakthrough products, services, or business processes that have the potential to disrupt or create entirely new markets. Substantial industry involvement in FTI projects, including financial, is mandatory to ensure quick market entry (within three years of the start of the project).

In 2018–2020, FTI had a total budget of $350 million (€300 million) with a call for proposals focused on the Horizon 2020 priority "Societal Challenges" and the specific objective "Leadership in Enabling and Industrial Technologies (LEITs)." Proposals are based on a business plan and focus on achieving high impact. FTI offers grants up to $3.5 million (€3 million) with funding levels fixed at 70% of the eligible costs.

5.4.6.3 Eureka/Eurostars

Eurostars is a joint program supporting start-ups in the research and development stage. It is the first European funding program dedicated to supporting innovative projects by start-ups targeting niche markets. With a bottom-up approach, it encourages and supports international collaborative projects with the potential to be rapidly commercialized.

5.4.7 Prizes

Horizon prizes of up to $11.6 million (€10 million) are part of the European Innovation Council pilot. They are challenge prizes or inducement prizes to whomever can most effectively meet a defined challenge. Prizes are open to a single person or legal entities, including EU-based start-ups, willing to think outside the box across sectors and disciplines.

These were the six challenges in 2018–2020:

1. "Innovative Batteries for Electric Vehicles"
2. "Fuel from the Sun: Artificial Photosynthesis"
3. "Early Warning for Epidemics"
4. "Blockchain for Social Good"
5. "Low-Cost Space Launch"
6. "Affordable High-Tech for Humanitarian Aid"

Challenges vary in relevance to the sustainable chemistry entrepreneur. In this particular set, the first is clearly the most relevant: making safe and sustainable batteries for electric vehicles through the development of new chemistries using low-cost, EU-abundant materials. Solutions must provide equal or better performance than vehicles with internal combustion engines and recharge in an equivalent time required to fill a conventional fuel tank. Expect additional challenges in future years to also require expertise in sustainable chemistry.

5.5 Setting Priorities When Pursuing Resources

As an innovator, you have limited time and money to spend on advancing your idea from the lab to the market. There simply isn't enough time to pursue every event, conference, grant proposal, or accelerator training. You likely have a finite amount of savings or income to pay for rent, travel to important meetings, and technical development. In lieu of money, it is also possible to give away too many ownership shares of your company if you engage too many for-profit accelerators or raise investment too early. You must prioritize.

These are some questions you can ask yourself to help prioritize which resources to pursue, and from whom:

- Networking events:
  Generally speaking, go! It's important to practice talking with as many different people as possible.
    o Is there a registration fee? How structured will it be?
    o Is the theme likely to yield new learnings or connections to relevant people?
    o Have I set aside time after the event to follow up with new connections?
- For practically *all other* resources:
    o Do I meet the eligibility criteria (such as location, stage of development, target industry, technologies they don't support, etc.)?
    o Is there anything that further makes my venture special given the criteria? (e.g. encouragement of women-led applicants if your venture is woman-led, having a

circular-economy business model when applying for a general cleantech competition, etc.)

- o Are there strings attached and am I comfortable with them?
- Grants:
  - o How long is the application/proposal? If I am not accepted, will I gain something valuable by going through the process anyway (e.g. refining the idea and getting feedback, incentive to get letters of support, ability to reuse parts in future proposals, etc.)?
  - o What are the expected milestones to be accomplished by the grant? Are they compatible with the business strategy, or am I just chasing the big money?
  - o What are the reporting requirements if I win the grant? How much time or money should I expect to dedicate to compliance?
  - o What is the timeline? How long does it take from proposal submission, to notification of acceptance/rejection, to disbursement of funds? (This should not be a barrier to applying but is good to know for budgeting purposes.)
- Competitions or prizes:
  - o If I don't win the grand prize, will I still gain something valuable as a result of participating (e.g. training, feedback, publicity, networking opportunities with relevant people, or other nonmonetary support)?
  - o Will I have to travel to a culminating event if selected as a finalist? If so, who pays for it/do I have funds to do so?
  - o What types of start-ups have typically won? Does my venture have characteristics in common that suggest I am a good contender?
- Accelerators and trainings:
  - o How long is the training and/or mentorship portion of the program?
  - o Is it virtual, or am I expected to be on-site for a residential program or other in-person events?
  - o How many hours per week are expected to be dedicated to in-class, advising, and homework activities related to the program?
  - o What is the curriculum? How do they match me with mentors or advisors?
  - o What have the outcomes been for teams who have gone through the program? What did they say was most and least valuable?
  - o Do I pay for my own travel and living expenses associated with it, or is there a grant/investment to defray costs?
  - o Is there a tuition fee that seems fair in exchange for what you will get?
  - o If the accelerator makes an up-front investment in exchange for a portion of ownership of my company, is it fair for what I am getting?
- Professional services, mentors, advisors:
  - o What are the expectations for the relationship (e.g. how often do you meet, how much do you need to brief them, is the advice specialized or general, etc.)?
  - o Who is referring them to me and why?
  - o What experience do they have with start-ups like mine? What onboarding and context-setting might I need to give them so that they can help me?

○ Is the advice confidential, or should I be careful to only speak generally? (This is especially important with investor-mentors and IP lawyers who may actually have a conflict of interest and will never sign confidentiality agreements in an informal context.)

○ If I am paying for time with these experts, how is payment structured (e.g. how much and how frequently)?

- Incubators with space:
  ○ How are resident companies selected for membership?
  ○ Are there tiers of membership? How much does it cost?
  ○ How is intellectual property protected? Does the incubator claim any IP rights? How do they ensure privacy, since companies usually have ongoing projects in shared labs or fabrication spaces?
  ○ What amenities are available – coffee, meeting rooms, special equipment, locker rooms, lab space? Do they cost extra?
  ○ To what extent do they offer educational activities, networking events, or match-making with important resources?
  ○ When do companies leave? Is there a trigger for leaving if they are under-performing or once they have grown too big for the space?

5.6   Conclusion: Engage with Your Ecosystem

It is critical for sustainable chemistry innovators to understand, explore, and engage with their ecosystems. As an entrepreneur getting to know your ecosystem, you will notice that most entrepreneurial resources do not explicitly support the innovation because it is chemistry or sustainability focused. Rather, you must sort through a lot of prospective opportunities and take responsibility to articulate how your innovation is aligned with each player's goals. By understanding the players and their motivations, you can more efficiently find, recognize the nuances, and make the case for support from different resources that can help you on your journey from lab to market to impact.

Entity offering resources	Entity usually funded by	Top reasons they support innovators	Types of resources usually offered	Scope	Level of venture development targeted
Government agency	Tax revenue	• Generate jobs and more tax-paying businesses • You will be a good corporate citizen that supports their tax base	• Funding: grants, sometimes investment or low-interest debt • Competitions/ prizes • Training	• Local • State/province • National	• Highly variable depending on program • Usually able to take more risks with teams still at bench-scale R&D phase

Entity offering resources	Entity usually funded by	Top reasons they support innovators	Types of resources usually offered	Scope	Level of venture development targeted
Nonprofit organization	• Government grants or contracts • Philanthropic/ corporate donations	• Alignment with mission • You are a part of their strategy to change the world	• Competitions and prizes • Acceleration training • Networks, coaching, and advisory services • Space • Grants, "friendly" investments	• Local • National • Global	• Highly variable depending on program • Usually able to take more risks with teams still at bench-scale or moving to small-scale production
Research Institutions • Universities • Government research centers	• Government grants • Testing fees • Revenues from technologies that have been licensed	• Mandate to ensure new technologies benefit society • You and your technology increase their reputation. You might also be a client to them or a source of licensing revenue	• Lab space • Third-party testing • Technical advisory services • Other non-profit-style resources	• Local • National	• Bench-scale, and translation from bench to small-scale production. • Pre-revenue start-ups spinning out of the university • Established companies seeking subcontracted research for scaling up to serve customers
Incubators	• For-profit: Rent or membership fees from start-ups; • Nonprofit: Grants or tax breaks from government	• Generate rent revenue for owners of the space • You are a paying client and a valued community member	• Infrastructure to house new businesses • Meeting and office space • Some offer lab space • Industry- specific networking events • A sense of community	• Local: you already live nearby • Global: you relocate	• Moving from bench to small-scale production • Have secured sufficient investment, grants, or self-funding to afford rent • Getting close to having sales

Entity offering resources	Entity usually funded by	Top reasons they support innovators	Types of resources usually offered	Scope	Level of venture development targeted
Accelerators • For-profit • Nonprofit/university	• For-profit: Investors such as wealthy individuals, larger investment groups, or industry-aligned corporations • See above for nonprofit	• Buy a share of the company while it is low-priced, cultivate it to increase value of shares • You are an investment with the expectation of bringing in follow-on investors	• Funding: Investment, sometimes a prize or grant • Training to prepare to raise more investment and gain customers • Specialized advisory and professional services • Office/lab space • Curated introductions to investors and customers • Cachet from affiliation with some programs	• Local • National • Global	• Usually after some initial training • Sufficient validation of the technology that a customer could pilot it • First investment, usually after you've won grants and prizes to get this far
Investors	• Angels: their own money • Venture capital groups: paid to manage their investors' money	• Alignment with investors' expertise and/or interest in a particular technology or industry • You are an investment. Your company is the "product" to eventually be sold to an acquirer	• Funding: investment • Managerial expertise and strategic advice • Industry connections	• Local: Angels • National or global: industry- or technology-specific VC firms	• Depending on the technology, they expect most R&D to be done and for their money to be used for sales growth

References

1 Griffen, T. (2017). 12 things about product-market fit. https://a16z.com/2017/02/18/12-things-about-product-market-fit (accessed 10 November 2020).

2 Van Roy, V., and Nepelski, D. (2017). Determinants of high-tech entrepreneurship in Europe. Joint Research Centre. https://core.ac.uk/download/pdf/84886731.pdf (accessed 30 October 2020).

3 Nepelski, D., Van Roy, V., and Pesole, A. (2019). The organisational and geographic diversity and innovation potential of EU-funded research networks. *The Journal of Technology Transfer* 44: 359–380.

4 Stam, E., and Spigel, B. (2016). Enterpreneurial ecosystems. Utrecht School of Economics working papers 16–13. https://ideas.repec.org/p/use/tkiwps/1613.html (accessed 30 October 2020).

5 Autio, E. (2016). Enterpreneuriship support in Europe: trends and challenges for EC policy. Technical report. doi: https://doi.org/10.13140/RG.2.1.1857.1762.

6

Factoring in Public Policy and Perception

Kira Matus

Division of Public Policy, HKUST, Clearwater Bay, Kowloon, Hong Kong

6.1 Introduction

Policy plays an important role in the development and use of chemicals generally, and those that are "green" or more sustainable in particular. Chapter 7 provides an overview of the legal regulatory requirements for bringing new chemicals to market in the US, the EU, China, and Canada. These regulatory frameworks are crucial for innovators to understand, as compliance is a necessary precondition for being able to sell new products.

Beyond these regulatory frameworks for new chemicals, there are a range of different policies, implemented by different parties, that can also impact the success of sustainable chemicals. Some of these are the result of particular government policies. These include green procurement policies for government agencies or the inclusion of municipal green building codes that reward specific, sustainable materials. Other regulatory frameworks have also emerged from nongovernmental sources. These include industry and third-party standards and certifications such as IFOAM Organic and OEKO-Tex®. It also includes purchasing requirements implemented by suppliers and retailers, such as Wal-Mart's 2005 decision to require that clamshell packaging in the US contain bio-based plastics [1]. Finally, there is a sort of "social regulation" that results from market and consumer preferences. These are often strongly influenced by public perceptions about the risks of specific chemicals.

All of these policies or regulations have two major implications for sustainable chemistry innovations. The first is that they influence the landscape as to which new products are realistically able to access specific markets. While a given chemical may be technically legal, the lack of a particular certification may make it difficult to penetrate a certain supply chain or retail outlet (see Chapter 8 for an overview of navigating supply chains). On the other hand, if it is favored by a particular standard, it may find a ready market.

The second major implication is the creation of market opportunities. As particular chemicals fall out of favor, opportunities arise for replacement by more sustainable alternatives. This is especially true for chemicals that have earned a reputation for being hazardous toward the environment or human health (BPA, PFOA/S, and brominated fire retardants being just a few examples). One of the major challenges for sustainable chemistry innovations is the strong staying power of incumbents [2, 3]. While governmental regulation often moves slowly when existing substances are suspected to be harmful [4], many certifications and the court of public opinion can move with greater speed. Being able to identify the opportunities presented by policy, regardless of its source, can be an important part of the success of a sustainable chemistry innovation.

This chapter will provide an overview of the different types of regulatory and policy environments that affect sustainable chemistry innovations and the potential impact on their commercialization.

6.2 Chemicals and Policy

The regulation of chemicals, to mitigate their impacts on human health and the environment, is one of the longest-standing environmental policy regimes [5]. Chapter 7 examines one important element of chemical policy and regulation, which is the requirement for the registration of chemicals new to the market. But there are many other different policies related to chemical management. These are implemented and enforced from the international level all the way down to local cities and towns. These policies include the following:

- Bans on the use of particular chemicals,
- Safety regulations for their use,
- Rules around storage, transport, and disposal,
- Air and water quality regulations,
- Pollution treatment policies, monitoring, and reporting.

Other policies may incentivize the use of particular chemicals through mechanisms such as mandatory alternative assessments or government preferred purchasing programs. While the breadth of these policies is beyond the scope of this chapter, this section will provide some examples of policies that are relevant for sustainable chemistry innovators in various geographic regions.

One way to classify these different policies is to divide them up between those that *directly target the chemical industry* and those that *indirectly target chemicals*. Both impact how chemicals are produced, used, and disposed. Examples of *direct* policies are those that require the registration (and sometimes testing) of new chemical substances. These include Toxic Substances Control Act (TSCA) in the United States and Registration, Evaluation, Authorisation and Restriction of Chemicals (REACH) in the EU, which are presented in Chapter 7. Other examples include policies that regulate the sale and use of pharmaceuticals, pesticides, and other chemicals such as hazardous waste. *Indirect* policies are often broader and include policies for general consumer protection, government green procurement requirements, and specific sets of standards and certifications (e.g. organic food standards, which dictate what sorts of pesticides farmers are able to use). Of course, this high-level division and categorization is rough, and there are often overlaps. But it is useful to understand that while some policies clearly target chemicals, there are others with impacts that are less obvious, at least initially. One important feature of chemical policies is that they exist at many different levels of government. These include international, national, and state/provincial agreements, as well local/municipal policies.

Providing a summary of all the chemical policies that exist, direct and indirect, is well beyond the scope of this chapter. There are several excellent resources that provide more details [5–7]. The rest of this chapter will be illustrative, as opposed to exhaustive, in its treatment of the importance of policy and regulation for sustainable chemistry innovation.

6.2.1 International Policies

There are several functions of international environmental policies, including those that focus on chemicals. The first is to provide harmonization. This prevents countries from "racing to the bottom," using lax environmental standards to give themselves an advantage over those that have more stringent requirements. These can create negative impacts on the environment and citizens of the less regulated country, as well as a variety of externalities that could spill over into neighboring regions. This is especially the case for pollutants that can travel long distances or have global impacts, like chlorofluorocarbons on the ozone layer. A second rationale is the fact that international policy divergence in the areas of health and the environment can constitute a significant barrier to the smooth functioning of economic markets. This is true for chemicals, where differences in hazard labeling, nomenclature, and handling recommendations can have significant environmental and safety repercussions. Finally, international policies may help to shift resources, especially knowledge, data, and expertise to areas where they are more limited.

For these reasons, chemicals are the subject of a number of international agreements. An excellent example is the 1987 Montreal Protocol on Substances that Deplete the Ozone Layer. This agreement regulates the consumption and use of more than 100 ozone-depleting substances [5, 6, 8]. It also incentivizes the development of more environmentally benign alternatives.

The work of the International Forum on Chemical Safety has led to several international treaties, including the Stockholm Convention (2001) on Persistent Organic Pollutants (POPs) and the Minamata Convention on Mercury (2013) [5, 6]. The Rio de Janeiro Earth Summit brought together multiple national governments, along with NGOs related to labor and economic cooperation, to begin to develop a globally harmonized chemical classification system (GHS). The GHS was established in 2003, which standardized chemical nomenclature, classification, and hazard labeling [6, 9].

Although international chemicals treaties have had the effect of reducing the trade and use of some hazardous chemicals, they have also understandably left gaps given there are approximately 100 000 chemicals currently in commerce [10]. There have been attempts, via the Strategic Approach to International Chemicals Management (SAICM), to promote chemical safety worldwide:

> SAICM's overall objective is the achievement of the sound management of chemicals throughout their life cycle so that by the year 2020, chemicals are produced and used in ways that minimize significant adverse impacts on the environment and human health. [11]

The SAICM laid out six different "Emerging Policy Issues" and two "Issues of Concern" (see Table 6.1) and has focused on collecting and disseminating information about these to

Table 6.1 SAICM issues identified for action as part of 2020 goals.

Emerging policy issues	Issues of concern
• Lead in paint	• Perfluorinated chemicals
• Chemicals in products	• Highly hazardous pesticides
• Electronic waste	
• Nanotechnology	
• Endocrine-disrupting chemicals	
• Pharmaceutical pollutants	

Source: Based on SAICM Emerging Policy Issues and Other Issues of Concern.

participating countries and stakeholders. However, it has yet to result in concrete changes to the international chemicals management regime [10]. The post-2020 framework is still under development. Despite this, the work of the SAICM is useful for sustainable chemistry innovators because it indicates areas where there is a degree of global consensus that there are problems with the current system. This provides demand for alternatives and therefore can act as a strong driver for innovation.

6.2.2 Regional Policy – The European Union

Arguably the biggest chemicals policy advance of the twenty-first century is the EU's REACH, which came into effect in 2007. REACH crystallized in policy a shift to a precautionary principle to chemicals management for a market that accounts for 16% of world exports and imports, and the world's second largest economy [12]. Concerns about the negative impacts of the US' approach since 1976 (discussed later) were a main driver of this fundamental change. REACH attempts to close the data gaps for existing chemicals and requires more testing for all chemicals, including those newly developed [6].

The European Commission's (EC) 2000 declaration states the following:

> It shall be based on the precautionary principle and on the principles that preventive action should be taken, that environmental damage should as a priority be rectified at source and that the polluter should pay. [13]

A key element of the design of REACH is to place the burden of proof on companies, as opposed to the regulator, to identify and manage hazards from chemical production and use [14]. This means that chemical registration with the European Chemical Agency (ECHA) requires testing data. Unlike in the US under TSCA, this data must be proactively provided. From this data, the ECHA has created a list of its "priority" chemicals that need to be assessed and has assigned them to different EU member states for further data collection and risk assessment. The member states report back to the ECHA with risk recommendations. For those that require regulation, the EC develops appropriate legislative proposals. (For more detail, see [15]).

As part of this shift of responsibility for data collection to companies, REACH also requires that companies work together to collect and submit the data for each chemical. The regulation specifies a broad range of "downstream user" companies with differing responsibilities depending on their categorization [16]. This includes making use of extended safety data sheets (eSDS) to check whether the intended use is already covered in the existing chemical registration and what, if any, measures are required for hazard reduction. The details of REACH are quite complex, but there are several important impacts of this program:

- The harmonization of chemical safety requirements across the EU, which constitutes more than 500 million people and approximately 22% of the world economy in 2019 (including the UK; as of writing, the status of the UK with regards to REACH after its departure from the EU is unclear) [17].
- A large and growing database of information containing hazard testing for chemicals used in commercial applications.
- A process of evaluation and authorization to identify and incentivize the development of less hazardous substitutes.

The practical implication of REACH for sustainable chemistry innovators (beyond the requirements in Chapter 7) is that REACH's substances of very high concern provide clear signals as to areas where there is a potential market for substitutes.

The ECHA also deals with EU-wide regulations on the import and export of hazardous materials, the Carcinogens and Mutagens Directive (CMD) to protect worker health and set limits for occupational exposure, the Waste Framework Directive to support the circular economy, and regulation that bans or restricts POPs throughout the EU [18].

The EU also has sector-specific regulations that are often relevant to chemical manufacturers, distributors, and users. These include the following:

- Cosmetics
- Restriction of hazardous substances in electrical and electronic equipment (RoHS)
- Food contact
- Construction products
- Safer detergents
- Biocidal products [18]

We will discuss these and similar sector- and product-specific regulations later in this chapter.

6.2.3 National-Level Policies

For countries outside of the EU, most chemicals policy is directed at the national level. The following sections describe the policies of selected countries and regions.

6.2.3.1 United States

The US' Toxics Substances Control Act (TSCA) was one of the first overarching chemicals policies when it come into force in 1976. Its goal, and that of other early chemical policies, was to create an inventory of chemicals in commerce. Prior to this point, it was largely unknown exactly which chemicals were actually in use. It also set up some of the earliest processes for the registration of new chemicals. However, in practice it became difficult for the US Environmental Protection Agency (EPA), which administers TSCA, to respond to the hazards presented by many chemicals in use. First, any chemical on the market prior to 1976 was exempt from authorization requirements, and the EPA could only compel review and require data submission if there was evidence of harm. This unfortunately created an incentive for companies to refrain from testing these substances since without evidence of impacts, the EPA was unable to act. Even for new chemicals, the testing requirements were minimal. TSCA was updated by the Lautenberg Act in 2016. The current state of the rules and procedures for new chemical registration are discussed in Chapter 7.

This is not the only set of regulations pertaining to chemicals. Pharmaceuticals must be tested and approved by the highly precautionary procedures of the Food and Drug Administration (FDA). While the EU has a reputation for taking a much more precautionary approach to chemicals broadly, the US has historically been much more cautious around the approval of drugs. In the late 1950s and early 1960s, doctors in the UK and Canada (and many other countries) prescribed thalidomide to pregnant women to help combat morning sickness. However, in the US, thalidomide was never approved for release onto the market due to an absence of test results. Within a few years, it became clear that thalidomide caused severe birth defects and was pulled from the market. This tragedy underscored the need for a rigorous testing regimen for new drugs, so the US model spread and was considered the "gold standard" for pharmaceutical regulation [19]. The stringent requirements of the FDA stand in contrast to those of TSCA, which requires little, if any, data for new chemicals not destined for use as pharmaceuticals.

There are also a number of policies that emerged between 1976 and 2016. These policies differed from TSCA in the sense that they targeted information provision and encouraged preventative actions to reduce risks arising from chemical use. The Pollution Prevention Act (1990) included the Toxics Release Inventory (TRI), which requires that firms report their emissions of particular chemicals at the plant/facility level as well as efforts taken to reduce these wastes. The goal was to reduce waste and improve the safety of waste management. This has proven to be an example of the power of measurement and reporting. The EPA estimates that actions reported as part of the TRI between 1991 and 2012 resulted in 390 000 different source reduction activities, with a total impact of between 2 and 6 million tons of toxic releases [20].

6.2.3.2 Canada

In Canada, chemical regulation falls under the Canadian Environmental Protection Act, 1999. All new chemicals produced in, or imported to, Canada since 1994 require assessment. Similar to the situation with TSCA, chemicals already in use are grandfathered out of these requirements. Their "categorization" program allowed them to identify 4300 priority chemicals and assess their risk and any need for regulatory action. The result of this exercise, in 2008, is the Chemicals Management Plan [21, 22]. As of 2016, the government was working to address the remaining 1500 chemicals on this list for potential action, with a goal of completing the assessment by 2020 [23].

6.2.3.3 Emerging Policies in East Asia: China and Taiwan

With the exception of Japan, chemicals policy in East Asia is relatively new and evolving. Given the sheer size of its chemical industry, the emerging regulatory structure in China is important in terms of both its impact on companies throughout the global chemical supply chain, as well as the Chinese population. The current regulation is the Measures for Environmental Administration of New Chemical Substances (MEP Order 7). In 2019, the Ministry of Ecology and Environment published a draft of a broad chemical regulatory program dubbed "China-REACH" by some, which was followed by revisions to the current regulation [24]. Generally speaking, the proposed changes would align China's program with TSCA and REACH [24]. This would include Chinese regulators collecting data from chemical companies to create a priority list of chemicals that are persistent, bio-accumulative or toxic (PBT), or very persistent and very bio-accumulative (vPvB). The government would monitor these substances and conduct risk assessments based on data provided by companies, leading to a variety of regulatory responses.

Taiwan issued its first high-level toxic and chemical substances regulation in 2019 [25]. The goal of the policy includes improved collaboration between the 13 different departments and agencies that are in some way involved with chemicals management and is based on the UN SAICM (see previous mention). Interestingly, the policy includes promotion activities for both sustainable chemistry and the circular economy as strategies to improve the efficiency of chemical use and mitigate the risks of chemical hazards to human health and the environment [25].

6.2.4 Policies Beneath the National Level (US)

In some cases, depending on the legal and political specifics, governments below the national level in the US are empowered to implement chemical policies within their jurisdictions. This can include specific requirements around hazardous waste handling and storage, restrictions (or even bans) on the sale or use of specific chemicals, and other control measures.

Starting around 2000, there has been a marked increase in the number of policies at the state and local levels, which has implications for manufacturers and users of chemicals. It is also providing opportunities for sustainable chemistry innovators, both in terms of direct support and market opportunities for sustainable alternatives. For example, Michigan established the Michigan Green Chemistry Program in 2006. This program included an annual Governor's Award program, a Green Chemistry Roundtable with approximately 20 members, and eight grants, between 2008 and 2011, that supported education, clean energy and advanced manufacturing, and green chemistry training workshops [26].

The State of California has one of the most comprehensive programs in place for the regulation of chemicals in consumer products and the promotion of green chemistry. The "California Green Chemistry Initiative" has existed since 2008 with two major goals: to accelerate the process of finding safe chemical alternatives and to create an online clearinghouse for information about toxic chemical substances [27]. California's Department of Toxic Substances Control (DTSC) has created a candidate list of chemicals of concern and of priority products that contain those chemicals. There are currently three combinations of chemicals and products that have been adopted for action, and another seven under consideration (see Table 6.2).

For the three adopted combinations of chemicals and products, manufacturers are required to submit an alternatives analysis, which evaluates the toxicity of the chemical of concern with alternative chemicals (or redesigns) to improve product safety. Once the

Table 6.2 California: priority products.

Adopted	Proposed
• Children's foam-padded sleeping products with Tris(1,3-dichloroisopropyl) phosphate (TDCPP) or Tris(2-chloroethyl) phosphate (TCEP) • Spray polyurethane foam with unreacted methylene diphenyl diisocyanate (MDI) • Paint or varnish paint strippers containing methylene chloride	• Carpets and rugs with per- and polyfluoroalkyl substances (PFASs) • Treatments containing PFASs on converted textiles or leathers • Laundry detergents containing nonylphenols and nonylphenol ethoxylates (NPE) surfactants • Paint and varnish strippers and graffiti removers containing N-methylpyrrolidone (NMP) • Nail products containing toluene • Nail products containing methyl methacrylate • Food packaging containing PFASs

Source: Priority Products, DTSC. © 2020 State of California.

DTSC evaluates these alternatives, it decides whether, and how, the product should be regulated. This can include requesting additional information, providing information to the public, use/sales restrictions, safety measures, end-of-life management, and targeted sustainable chemistry/engineering innovations to produce feasible alternatives [28]. If the manufacturer fails to comply, importers, distributors, and/or retailers of the product are responsible for the submission.

6.3 New Trends and Approaches

Since the first generation of comprehensive chemical policies at the start of the modern environmental regulatory era in the 1970s and 80s, there has been a change in the nature of the challenges facing policy-makers and decision-makers. In the 1970s, one of the major challenges was not knowing the full range of chemicals used in commercial enterprises. From that perspective, early policy frameworks in the US, Europe, Japan, and elsewhere successfully solved the initial information gap issue in terms of "what is there." These policies were less successful at incentivizing R&D focused on existing or new chemicals with reduced impact on health and the environment.

The concern over impacts has grown with advances in toxicology and epidemiology that have overturned old models of "safe" thresholds for certain chemicals. Carcinogens have long been a matter of public concern, but these concerns have expanded over the past two decades to other chemical "threats," especially toxins that bioaccumulate in populations or the environment and/or disrupt endocrine function. Bioaccumulative chemicals, even in miniscule amounts over time, could have harmful impacts. Exposure to endocrine disruptors in the womb could have detrimental effects lasting for multiple generations [29, 30].

6.3.1 The Precautionary Shift

This precautionary shift in the chemicals "problem" has been at least a partial driver behind trends in new approaches to chemicals policy. The US' Pollution Prevention Act was part of a larger trend toward pollution prevention, as opposed to "end of pipe" controls. Green chemistry, as elucidated by the Twelve Principles of Green Chemistry by Warner and Anastas in 1998 [31], is itself a direct outgrowth of this evolving preventative approach to hazard reduction.

As described by Hansen et al. [7], policy experts have identified these four concrete elements that constitute the application of the precautionary principle:

- Taking preventive action in the face of uncertainty
- Shifting the burden of proof or responsibility to proponents of potentially harmful activities
- Exploring a wide range of alternatives to possibly harmful actions
- Increasing public participation in decision-making

The incorporation of some combination of these elements into policy can be seen at multiple levels around the world.

As discussed earlier, the EU started developing a regulatory system for chemicals explicitly based on a more precautionary approach in 2000. Though negotiations between member states and also with industrial stakeholders weakened the final legislation compared to the initial proposals (in terms of the strength of application of the principle), it was still a distinct innovation over existing approaches at that time. While it does not cover all four elements, the shift of the burden of proof onto the chemical industry and the requirement for alternatives assessment for chemicals of concern are precautionary elements that feature prominently in REACH [32].

The trend toward a precautionary approach has not been limited to the EU. While there was little (if any) evidence of a precautionary approach in the final legislation that updated TSCA in the US [32], an analysis of local chemical policies in the US reveals explicit declarations in support of the precautionary principle at the state and municipal levels. For example, San Francisco passed a Precautionary Principle Ordinance in 2003, which states the following:

> All officers, boards, commission, and departments of the City and County shall implement the Precautionary Principle in conducting the City and County's affairs: The Precautionary Principle requires a thorough exploration and a careful analysis of a wide range of alternatives. Based on the best available science, the Precautionary Principle requires the selection of the alternative that presents the least potential threat to human health and the City's natural systems. Public participation and an open and transparent decision-making process are critical to finding and selecting alternatives [33].

Some policy makers have also been evaluating ways to implement chemicals policies that consider broader classes of chemicals. This is to reduce the propensity to regulate chemical by chemical, often in response to public concern about "the baddie of the week."

6.3.2 Attention to Vulnerable Populations

Another shift in chemicals policy has been an increased attention to vulnerable populations. For example, there have been several scandals about toxics found in children's products, including melamine in Chinese milk powder, lead and cadmium in paint on toys, brominated flame retardants on children's pajamas, and BPA in baby bottles. These have led to a series of approaches to chemicals management that requires more attention be paid to groups. These include fetuses, babies, children, and the elderly who may suffer more adverse consequences from exposure to certain chemical substances [34]. Citizens' groups representing parents in the US have lobbied and partnered with NGOs resulting in state- and national-level policies that update chemical regulation [35–37]; the Lautenberg Act specifically requires that the US EPA take action to identify and protect susceptible groups, including pregnant women, infants, children, and the elderly [38].

In the EU, REACH has general statements about the protection of vulnerable groups (although definitions of "vulnerable groups" are generally vague) [39]. This has been flagged as an area that needs to be addressed across a number of EU regulations that relate to chemical hazards [39]. In Canada, Health Canada and Environment and Climate Change Canada undertook a consultation in 2018/2019 on a proposed definition of "vulnerable populations" to inform chemical management actions by the federal government [40].

As part of their Chemicals Management Plan, the Canadian government has a human biomonitoring program that, among other things, is meant to help identify groups that have been exposed to substances of concern at higher levels. There are also targeted biomonitoring initiatives that include environmental lead levels, long-term residential air pollution exposure in children, plastics and personal care products use in pregnancy, and exposure to acrylamide through vulnerable populations' consumption of contaminated food [22]. These are just a few examples of developing programs that are furthering understanding, and potentially policy action that focuses on vulnerable groups.

6.3.3 Industry, NGOs, the Public, and Other "Governance" Actors

One major broad change in environmental policy that is reflected in chemicals policy is a move toward "governance." This refers to "governing without government" [41]. Chemical "governance" involves policies and regulations that have emerged separate from the familiar, formal mechanisms of governments. Reflecting a growing demand from the public to reduce exposure to chemicals of concern, these include regulations within firms, such as Walmart's Sustainable Chemistry Program, industry-level initiatives like Responsible Care®, and a growing number of third-party standards and certifications.

Consumers in the marketplace are demanding transparency and information on substances found in the products they purchase. This has led to programs such as GoodGuide® – a ranking of more than 75 000 products based on their contents [42]. These various standards and certifications lead the charge ahead of many government-based programs. They can also become de facto gateways to market access, even in the absence of traditional regulation. Furthermore, these governance activities can at times act as a forerunner, becoming the basis for legally binding government action.

An example of this is green building standards. Many elements of Leadership in Energy and Environmental Design (LEED), which includes rules about the use of specific materials like low VOC paints, and its international equivalents have been incorporated in building codes worldwide [43]. This has created space for a number of different sustainable innovations, including within the construction chemicals space (e.g. adhesives, bonding agents, mortars). The Substitute It Now (SIN) List is a database of chemicals likely to be banned, based on REACH [42]. Information such as the SIN list, the GoodGuide, and the Chemical Footprint Project put pressure on firms to move beyond their strict legal responsibilities in response to market pressure. These efforts create incentives, in the form of market creation, for a wide range of more sustainable alternatives and competitive innovations.

6.3.4 Public Perceptions

The preferences and decisions of the public are one factor that drive the demand for public policy and governance initiatives that address chemical hazards. Public perception of chemicals has been evolving in interesting and sometimes challenging ways, especially for incumbent products that are perceived to be harmful. Modern society has undoubtedly reaped huge benefits from the tens of thousands of chemicals that have entered commerce over the past century, and most of the time, most people rarely consider the role of chemicals in their daily lives. At the same time, many chemicals are particularly hazardous. Of these hazardous chemicals, a relatively small number have become targets of controversy and even downright fear – and we have only a limited understanding of why the public deem certain chemical substances to be "safe" or "dangerous."

This has led to behaviors that are puzzling to many chemists, who are much more fluent with the concepts of chemical hazard and risk. For example, there are numerous companies marketing personal care or cleaning products as "chemical free" – clearly an impossibility. In the market, there are many cases where "naturally based" or "green" household cleaners command a price premium. This "green premium" is discussed in greater detail in Chapter 3. There are also large classes of chemical products (e.g. plastics or pesticides) that have found widespread use and yet have suddenly become the subject of public concern and increased regulation. These regulations and concerns are at least one driving factor of the continued growth in development and sales of bioplastics (plastics that either are sourced from renewable materials, such as corn, and/or are biodegradable at the end of life). While bioplastics currently comprise less than 1% of the plastics market, recent market analysis predicts demand growing at a rate of 20–30% per year. This is in addition to demand for the development of new additives alongside this increasing share of the plastics market [44, 45].

It is clear that the public has a complex perception of the risks posed by different kinds of chemicals, and perhaps even what the word "chemical" means. We see reactions to some chemicals by individuals, markets, and governments, including overly hasty shifts to equally or differently problematic substances. Similarly, policies considered burdensome and reactionary by some will not be nearly stringent enough for others.

We have little understanding of individuals' attitudes toward chemical hazards when compared or ranked against other risks that they encounter. The phenomenon of "chemophobia," the irrational fear of chemicals, has been studied from the perspective of stigmatization and risk perception in an effort to understand its roots and to prevent harm associated with it [46]. Studies have linked chemophobia to the rejection of beneficial products, such as vaccines [47], and to self-endangerment from the mishandling of "eco-labeled" consumer products erroneously thought to be a safer substitute [46].

This fear is emblematic of a misalignment between "expert" and "lay" understanding of the sources of chemical risk. The polarity of opinions and responses to the same "chemicals category" highlights the need for further research on the determinants of willingness to accept risk and how that relates to differing attitudes regarding the need for government intervention.

This gap in our understanding proves problematic for policymaking and for businesses. Addressing chemical hazards to protect the public implies that policy and decision-makers need to account for their own understanding and bias, for scientific evidence and advice, and for public concern and demand for increased regulation. At the same time, they're faced with pressure to balance investment and the development of an industry that not only has steadily improved our quality of life and life expectancy, but also generates trillions of dollars annually worldwide [48]. It poses a challenge for businesses that may find their products to be suddenly considered harmful after years on the market or that may not even be fully aware of the chemical composition of the goods that they are retailing.

6.4 Conclusion: Policy as Strategic Advantage for the Sustainable Chemistry Innovator

6.4.1 Perceptions and Opportunities

Early perceptions of the relationship between environmental regulation and innovation tended to take a negative view. It was implied that forcing firms to comply with various environmental, health, and safety regulations meant they would have to redirect funds that would otherwise have been directed toward innovative R&D [49]. Further work has shown that this is not always the case. In fact, stringent environmental regulation can positively impact innovation, often by making existing products and processes too costly and in need of replacement [49, 50].

For the sustainable chemistry innovator, chemical policy can provide crucial market intelligence and opportunities. Despite years of discussion of the potential of "green chemistry" to help transform the chemical industry, and the attendant predicted financial opportunities, actual implementation remains challenging [51]. Yet according to one industry group, the Green Chemistry & Commerce Council (GC3), it is regulation and consumer awareness that drive the uptake of sustainable chemistry innovations in industry [51, 52]. For example, there is a correlation between REACH regulations and increased patent filings of nonphthalate inventions. Similarly, action by the California DTSC has increased the demand for nonphthalate plasticizers [52].

This has important strategic implications for innovators. Many of the national policies described in this chapter consist of lists of chemicals of concern, and national and regional plans to address them in groups over time. This is a major signal to innovators of areas where more sustainable alternatives have a large potential to unseat incumbent products and processes. Market sectors where there are burgeoning standards and certifications, and public pressure for transparency, will be particularly receptive to new technologies that can legitimately claim to reduce or eliminate specific hazards. Policy may indeed become a burden for existing products and processes that are less socially acceptable and more expensive, while providing important commercial avenues for alternative sustainable chemistry technologies.

6.4.2 Practical Actions

There are several key messages and actions that sustainable chemistry innovators should consider as you define your technology and product and build your business.

- Understand the relevant policy environment of the country and region where you are commercializing your technology. This includes both the legal regulatory framework discussed in Chapter 7 as well as the public policy landscape addressed here.
- Determine whether/how your product fits into these frameworks.
- Pay attention to areas of concern highlighted by NGOs and the media. These can result in market opportunities for new sustainable alternatives, even in advance of regulation.
- Emphasize any potential benefit due to the policy environment to maximize the competitive advantage of your product. Make sure you can put this in layperson terms.
- Consider whether your product would benefit by engaging in certification programs. Some of these can be costly but may be a useful way to access new market segments, especially for sustainability-focused buyers.

Acknowledgments

Thank you to Marie Bernal for her support, especially in combing through the public perceptions literature.

References

1 Matus, K.J.M. (2010). Innovating for development: policy incentives for a cleaner supply chain: the case of green chemistry. *Journal of International Affairs* 64 (1): 121–136.

2 Matus, K.J.M., Clark, W.C., Anastas, P.T., and Zimmerman, J.B. (2012). Barriers to the implementation of green chemistry in the United States. *Environmental Science & Technology* 46 (20): 10892–10899. https://doi.org/10.1021/es3021777.

3 Matus, K.J.M., Xiao, X., and Zimmerman, J.B. (2012). Green chemistry and green engineering in China: drivers, policies and barriers to innovation. *Journal of Cleaner Production* 32: 193–203. https://doi.org/10.1016/j.jclepro.2012.03.033.

4 Krimsky, S. (2017). The unsteady state and inertia of chemical regulation under the US toxic substances control act. *PLoS Biology* 15 (12): e2002404. https://doi.org/10.1371/journal.pbio.2002404.

5 Selin, H. (2010). *Global Governance of Hazardous Chemicals: Challenges of Multilevel Management*, Politics, Science, and the Environment. Cambridge, Mass: MIT Press.

6 Geiser, K. (2015). *Chemicals without Harm: Policies for a Sustainable World*, Unban and Industrial Environments. Cambridge, Massachusetts: The MIT Press.

7 Bergeson, L.L. and American Bar Association (eds.) (2014). *Global Chemical Control Handbook: A Guide to Chemical Management Programs*, 1e. Chicago, Illinois: American Bar Association, Section of Environment Energy, and Resources.

8 UN Environment. (n.d.). About Montreal protocol. https://www.unenvironment.org/ozonaction/who-we-are/about-montreal-protocol (accessed 9 July 2020).

9 McEldowney, S. (2017). Sustainable chemical regulation in a global environment. In: *Natural Resources and Sustainable Development*, 257–277. Edward Elgar Publishing https://doi.org/10.4337/9781783478385.00019.

10 Backhaus, T., Scheringer, M., and Wang, Z. (2018). Developing SAICM into a framework for the international governance of chemicals throughout their lifecycle: looking beyond 2020. *Integrated Environmental Assessment and Management* 14 (4): 432–433. https://doi.org/10.1002/ieam.4052.

11 SAICM. (n.d.). SAICM overview. http://www.saicm.org/About/SAICMOverview/tabid/5522/language/en-US/Default.aspx (accessed 9 July 2020).

12 European Commission. (2019). EU position in world trade - Trade - European Commission. https://ec.europa.eu/trade/policy/eu-position-in-world-trade (accessed 27 June 2020).

13 Consolidated Versions of the Treaty on European Union and the Treaty of the Functioning of the European Union 2016. Vol. 2016/C 202/01, p Article 191, paragraph 2.

14 ECHA. (n.d.). Understanding REACH. https://echa.europa.eu/regulations/reach/understanding-reach (accessed 9 July 2020).

15 Heyvaert, V. (2010). Regulating chemical risk: REACH in a global governance perspective. *Regulating Chemical Risks*: 217–237.

16 ChemicalWatch. (2018). Regulatory impact report: the impact of reach on downstream users. https://chemical-watch.s3.amazonaws.com/downloads/Chemical-Watch_Regulatory-Impact-Report.pdf (accessed 10 November 2020).

17 IMF. (n.d.). Report for selected country groups and subjects. https://www.imf.org/external/pubs/ft/weo/2016/02/weodata/weorept.aspx?pr.x=56&pr.y=8&sy=2016&ey=2016&scsm=1&ssd=1&sort=country&ds=.&br=1&c=001,998&s=NGDPD&grp=1&a=1 (accessed 9 July 2020).

18 ECHA. (n.d.). Legislation. https://echa.europa.eu/legislation (accessed 9 July 2020).

19 Vogel, D. (2012). *The Politics of Precaution: Regulating Health, Safety, and Environmental Risks in Europe and the United States.* Princeton University Press.

20 Ranson, M., Cox, B., Keenan, C., and Teitelbaum, D. (2015). The impact of pollution prevention on toxic environmental releases from U.S. manufacturing facilities. *Environmental Science & Technology* 49 (21): 12951–12957. https://doi.org/10.1021/acs.est.5b02367.

21 Government of Canada. (n.d.). Canada's system for addressing chemicals. https://www.canada.ca/en/health-canada/services/chemical-substances/canada-approach-chemicals/canada-system-addressing-chemicals.html (accessed 9 July 2020).

22 Government of Canada. (n.d.). Monitoring and surveillance activities under Canada's Chemicals Management Plan. https://www.canada.ca/en/health-canada/services/chemical-substances/chemicals-management-plan/monitoring-surveillance.html (accessed 9 July 2020).

23 Government of Canada. (n.d.). Chemicals Management Plan. https://www.canada.ca/en/health-canada/services/chemical-substances/chemicals-management-plan.html (accessed 9 July 2020).

24 ChemicalWatch. (2019). Expert Focus: China's evolving chemical regulatory regime – implications for businesses https://chemicalwatch.com/80926/expert-focuschinas-evolving-chemical-regulatory-regime-implications-for-businesses#overlay-strip (accessed 9 July 2020).

25 ChemicalWatch. (2018). Taiwan approves 'national policy plan' to manage chemicals. https://chemicalwatch.com/asiahub/67657/taiwan-approves-national-policy-plan-to-manage-chemicals (accessed 9 July 2020).

26 Michigan Department of Natural Resources and Environment. (2010). DEQ - Green Chemistry. http://www.michigan.gov/deq/0,1607,7-135-3585_49005---,00.html (accessed 28 August 2010)

27 DTSC. (n.d.). How are the Safer Consumer Products Regulations related to the Green Chemistry Law? https://dtsc.ca.gov/how-are-the-safer-consumer-products-regulations-related-to-the-green-chemistry-law (accessed 9 July 2020).

28 California Department of Toxic Substances. (2018). Three year priority work plan 2018–2020. Safer Consumer Products Branch. p 29.

29 Myers, J.P., Zoeller, R.T., and Vom Saal, F.S. (2009). A clash of old and new scientific concepts in toxicity, with important implications for public health. *Environmental Health Perspectives* 117 (11): 1652.

30 vom Saal, F.S. and Myers, J.P. (2008). Bisphenol A and risk of metabolic disorders. *JAMA: The Journal of the American Medical Association* 300 (11): 1353–1355.

31 Anastas, P.T. and Warner, J.C. (1998). *Green Chemistry : Theory and Practice*. Oxford England; New York: Oxford University Press.

32 Hansen, S.F., Carlsen, L., and Tickner, J.A. (2007). Chemicals regulation and precaution: does REACH really incorporate the precautionary principle. *Environmental Science & Policy* 10 (5): 395–404.

33 San Francisco Env. Code 2003.

34 Kira J M Matus; Marie N Bernal. (1998). *Media Attention and Policy Response: 21st Century Chemical Regulation in the USA*. Science and Public Policy.

35 Belliveau, M.E. (2011). The drive for a safer chemicals policy in the United States. *NEW SOLUTIONS: A Journal of Environmental and Occupational Health Policy* 21 (3): 359–386. https://doi.org/10.2190/NS.21.3.e.

36 Tickner, J., Geiser, K., and Coffin, M. (2005). The US experience in promoting sustainable chemistry (9 Pp). *Environmental Science and Pollution Research* 12 (2): 115–123.

37 Tickner, J.A. and Geiser, K. (2004). The precautionary principle stimulus for solutions- and alternatives-based environmental policy. *Environmental Impact Assessment Review* 24 (7–8): 801–824. https://doi.org/10.1016/j.eiar.2004.06.007.

38 Koman, P.D., Singla, V., Lam, J., and Woodruff, T.J. (2019). Population susceptibility: a vital consideration in chemical risk evaluation under the Lautenberg toxic substances control act. *PLoS Biology* 17 (8): e3000372–e3000372. https://doi.org/10.1371/journal.pbio.3000372.

39 Kuipers, W. Y.; Mascolo, M. (2017). *Study for the Strategy for a Non-Toxic Environment of the 7th EAP: Sub-Study: Protection of Children and Vulnerable Groups from Harmful Exposure to Chemicals*. European Commission Directorate-General for Environment Sustainable Chemicals. p. 126.

40 Government of Canada. (2019).What we heard: Defining vulnerable populations. https://www.canada.ca/en/health-canada/services/chemical-substances/consulting-future-chemicals-management-canada/what-we-heard-defining-vulnerable-populations.html (accessed 9 July 2020).

41 Kersbergen, K.V. and Waarden, F.V. (2004). 'Governance' as a bridge between disciplines: cross-disciplinary inspiration regarding shifts in governance and problems of governability, accountability and legitimacy. *European Journal of Political Research* 43 (2): 143–171. https://doi.org/10.1111/j.1475-6765.2004.00149.x.

42 Swarr, T.E., Cucciniello, R., and Cespi, D. (2019). Environmental certifications and programs roadmap for a sustainable chemical industry. *Green Chemistry* 21 (3): 375–380. https://doi.org/10.1039/C8GC03164A.

43 Donadelli, F. and Matus, K. (2020). Using Private Regulation for the Public Good. In: *The Palgrave Handbook of the Public Servant* (eds. H. Sullivan, H. Dickinson and H. Henderson), 1–15. Cham: Springer International Publishing https://doi.org/10.1007/978-3-030-03008-7_6-1.

44 GlobeNewswire. (2019). Bioplastic packaging market will grow at CAGR of 27.6% to hit $28.57 billion by 2025. https://www.globenewswire.com/news-release/2019/04/25/1809278/0/en/Bioplastic-Packaging-Market-will-grow-at-CAGR-of-27-6-to-hit-28-57-Billion-by-2025-Global-Analysis-by-Price-Trends-Size-Share-Investment-Opportunities-and-Key-Players-Adroit-Market.html (accessed 30 October 2020).

45 GlobeNewswire. (2020). Global market for bioplastics (2020 to 2030) - featuring Aquafil, Arkema & BASF Among Others. https://finance.yahoo.com/news/global-market-bioplastics-2020-2030-102856811.html (accessed 30 October 2020).

46 Saleh, R., Bearth, A., and Siegrist, M. (2019). "Chemophobia" today: consumers' knowledge and perceptions of chemicals. *Risk Analysis* https://doi.org/10.1111/risa.13375.

47 Entine, J. (2011). *Scared to death: how chemophobia threatens public health: a position statement of The American Council on Science and Health*. American Council on Science.

48 Oxford Economics. (2019). *The Global Chemical Industry: Catalyzing Growth and Addressing Our World's Sustainability Challenges*. ICCA.

49 Acar, O.A., Tarakci, M., and van Knippenberg, D. (2019). Creativity and innovation under constraints: a cross-disciplinary integrative review. *Journal of Management* 45 (1): 96–121. https://doi.org/10.1177/0149206318805832.

50 McEntaggart, K.; Etienne, J.; Beaujet, H.; Campbell, L.; Blind, K.; Ahmad, A.; Brass, I. (2020). *Taxonomy of Regulatory Types and Their Impacts on Innovation*. BEIS Research Paper Series. No. 2020/004. BEIS.

51 Veleva, V.R. and Cue, B.W. (2019). The role of drivers, barriers, and opportunities of green chemistry adoption in the major world markets. *Current Opinion in Green and Sustainable Chemistry* 19: 30–36. https://doi.org/10.1016/j.cogsc.2019.05.001.

52 Green Chemistry & Commerce Council (GC3). (2015). *Advancing Green Chemistry: Barriers to Adoption and Ways to Accelerate Green Chemistry in Supply Chains*. GC3.

7

Pre-market Approval of Chemical Substances: How New Chemical Products Are Regulated

Richard E. Engler

Bergeson & Campbell, P.C. and The Acta Group, Washington, DC, USA

7.1 Introduction

Other chapters in this book provide a wide variety of information for chemical innovators looking to bring a new technology to market: how to protect intellectual property, develop partnerships, raise funding, and generate the product in a capital-intensive industry. This chapter focuses on a more under-appreciated aspect of commercialization that is specific to chemical substances: pre-market review and approval.

It is generally well-known that pharmaceuticals undergo extensive rounds of testing. Most innovators probably understand the meaning of and need for clinical trials to ensure product safety. There is far less appreciation for what is required for chemicals being commercialized for other uses. While such an understanding may have limited utility for academic researchers, any innovator looking to advance a technology to market needs to at least be aware of the necessary regulatory compliance steps. By understanding the requirements and planning ahead, innovators can pave the way for the successful commercial launch of their product.

Note: This chapter is intended to give a high-level view of what is required to commercialize chemical products. It is not to be construed as legal advice. Nor does this chapter address all the details of testing obligations, pre-market review, or ongoing obligations. A chemical supplier (manufacturer or importer) must ensure that it has robust policies and procedures to ensure compliance before and after market entry. The complete details of such obligations are far beyond the scope of this chapter. Incidence of noncompliance can range from minor to severe. Even if there is no harm to humans or the environment from an incident of noncompliance, the commercial disruptions that result from having to cease manufacture while returning to compliance can be as damaging, if not more so, to a business than the actual fine imposed. For a small, early-stage company, the result can be devastating.

Estimates of how long/how much to get to market are designed to give a *rough* idea of the time and expense to obtain pre-market approval (in cases where it is indeed granted) under each use case. Costs may be lower if there is existing data that can be leveraged to support the assessment of the product. Costs may be substantially higher if a product is novel, leading a regulator to request or require additional time and testing, or if testing is especially difficult,

How to Commercialize Chemical Technologies for a Sustainable Future, First Edition.
Edited by Timothy J. Clark and Andrew S. Pasternak.

requiring additional method development by the testing lab. Cost estimates include testing and government fees but exclude costs of consultants or external counsel.

7.2 Overview

The regulation of chemical products (as opposed to the regulation of pollution or waste) is complex. It varies depending on the country or region and the intended use of the chemical product. In this chapter, we will discuss at a high level the regulations that apply to placing chemical substances on the commercial market in four major markets: the US, the EU, China, and Canada. More detailed information on each of these markets is available in the *Global Chemical Control Handbook: A Guide to Chemical Management Programs* [1]. It is critical that companies evaluate carefully the requirements for each use in each jurisdiction – the obligations that arise are very fact-specific.

In general, chemical regulations divide products into two categories: *chemical substances* and *articles*. While there are differences among regulatory regimes about the precise definition of an article, an article is generally considered to be a product that is manufactured with a specific design or function, and the end use is dependent on that shape or design. Gases, liquids, and powders are unlikely to meet the definition of an article because these products do not have a specific shape or design. While articles themselves are exempt from most chemical control regulations, substances that are intentionally released from an article are not. For example, a toner cartridge is an article, while the toner is not. An article may be simple (e.g. a wire) or complex (e.g. an airplane) and may consist of one substance or a combination of substances. If a product is not an article, it is a substance or a mixture of substances. Here are the definitions for the purposes of this chapter:

- Product: Something that is imported or manufactured. It includes articles, substances, and mixtures.
- Article: A product with a specific design or function where the end use of the product depends on the shape or design.
- Substance: A chemical substance (in any physical state) that is not an article.
- Mixture: A combination of substances (in any physical state) that do not undergo a chemical reaction.

Regulatory obligations flow from the intended use of the product. Pharmaceuticals generally face the most thorough review: tests for safety and efficacy. Food (including animal feed) and food additives are also carefully scrutinized for safety. Any product that is intended to be introduced into, or consumed by, humans rightly requires the highest level of review. Pesticides also face significant pre-market review – any product that is intended to kill, control, or mitigate other living organisms may have collateral effects on nontarget species or the people using those products. Food contact substances and substances used in medical devices [2] also deserve special consideration as either may unintentionally introduce undesired, deleterious substances to humans. Cosmetics must be demonstrated as safe for use as intended, although not all countries require that a regulatory authority review the finding. Other uses are regulated as "industrial" chemicals, although this term usually includes commercial and consumer uses. If a substance is intended for more than one of these use categories, the manufacturer would have to satisfy the regulatory requirement for the substance in each of those use categories.

A chemical substance is often associated with a numeric identifier, such as a Chemical Abstracts Service (CAS) registry number (RN) or European Community (EC) number. On the other hand, regulatory chemical nomenclature is not as simple as writing an International Union of Pure and Applied Chemistry (IUPAC) name, checking an SDS authored by a specialty chemical company, or finding an identity on a list and assuming that it is "close enough." Be sure to establish the correct identity for a product for a particular use in a particular country or region. Without the correct substance identity, a manufacturer may not be able to establish whether the substance is allowed for that use, allowed with some restrictions, or a new substance requiring some pre-market notice and/or approval.

A substance may be well-defined, such as ethanol (CAS RN 64-17-5), which can be represented by a single structure, have a range of molecular weights such as polyethylene (*ethene, homopolymer*, CAS RN 9002-88-4), or be a complex combination of a number of substances (an unknown, variable, complex, or biological substance or UVCB) such as sandalwood oil (*oils, sandalwood;* defined as: *Extractives and their physically modified derivatives. Santalum album, Santalaceae*, CAS RN 8006-87-9). Although it may be difficult for a chemist to understand, how a particular substance is defined depends on the use and the country or region. Innovators must be mindful of the particular nomenclature rules for the target jurisdiction.

In each country or region discussed, only a legal entity within the country may seek pre-market approval. Canada and China allow a foreign supplier to appoint an agent that acts on the supplier's behalf. In the EU, a company can appoint an "Only Representative" – a legal entity in the EU that meets all the requirements for a chemical registration. In the US, the domestic customer of a foreign supplier is normally the entity that satisfies the regulatory requirements, although the foreign supplier is often intimately involved, especially in developing and providing necessary information to support pre-market approval.

7.3 United States

The two agencies in the US that oversee the safety of the vast majority of chemical products are the US Food and Drug Administration (FDA) and the US Environmental Protection Agency (EPA). The FDA administers the Federal Food Drug and Cosmetic Act (FFDCA). The EPA administers Federal Insecticide, Fungicide, and Rodenticide Act (FIFRA) and the Toxic Substances Control Act (TSCA).

FFDCA, FIFRA, and TSCA all recognize the importance of research and development (R&D), and all have provisions to allow for some manufacture of substances for R&D purposes. R&D is expected to be performed under the auspices of a qualified individual, and distribution is limited to others that are performing R&D. R&D exemptions often include provisions for scale-up, but the innovator should seek regulatory guidance prior to beginning clinical trials or pesticide efficacy or field testing.

7.3.1 Federal Food Drug and Cosmetic Act (FFDCA)

FFDCA is the statute that oversees pharmaceuticals, medical devices, food (and feed), food additives and food contact substances, and cosmetics. The requirements for each use category vary. In nearly all cases, the regulatory burden falls upon the manufacturer (or importer) of the final product, whether it be the dosage form, medical device, food or feed additive, food contact material, or cosmetic formulation. Nearly all chemicals that are intended to be used in an FDA-regulated use must be manufactured under Good Manufacturing Practices (GMP). Under FFDCA, the manufacturer of the end product has the legal responsibility to ensure that the product is safe and effective when used as intended and that the product is not adulterated or misbranded. If a substance is supplied to an end-product manufacturer, that manufacturer may require the substance supplier to provide the data necessary for the end-product manufacturer to document its compliance with the appropriate regulations.

7.3.1.1 Pharmaceuticals

Drugs are defined as products that are "intended for use in the diagnosis, cure, mitigation, treatment, or prevention of disease" and "(other than food) intended to affect the structure or any function of the body of man or other animals" [3].

Contrast this with the definition of a cosmetic as products "intended to be rubbed, poured, sprinkled, or sprayed on, introduced into, or otherwise applied to the human body…for cleansing, beautifying, promoting attractiveness, or altering the appearance" [3].

Often the difference comes down to statements or "claims" made for the product. If the product is claimed to "heal skin," it is likely a drug, not a cosmetic. If, on the other hand, the product is claimed to "improve the appearance of skin," it is likely a cosmetic, not a drug. This distinction shines a light on some of the subtleties that might be lost on an innovator more accustomed to scientific issues.

Components of drugs, including intermediates, catalysts, and other substances used to manufacture ingredients of the final dosage form, also fall under FFDCA jurisdiction and generally do not undergo pre-market review; rather, such substances are evaluated in the review of the final drug and whether the substances are present in the final form. Once a drug is approved, there are limited opportunities to change the manufacturing process absent some level of re-review by the FDA, so it is *important to incorporate sustainability (e.g. use of a green solvent or catalyst rather than stoichiometric reagents for the synthesis) as early in the drug development process as possible.*

How long/how much? Regulatory approval for a new drug is extremely time- and resource-intensive. Innovators and investors should expect testing (mammalian and human clinical trials) and review and approval by the FDA to take years and cost millions of dollars. Some of the regulatory process can run in parallel to process development and scale up, but the safety assessment must be done on the final form, so the manufacturing process needs to be set fairly early in the regulatory approval process. Changes in a manufacturing process to incorporate more sustainable chemistry do not necessarily require an entirely new set of clinical testing, but the safety profile of the final drug product will have to be re-reviewed by the FDA. If there are differences in the composition of the final product (including impurities), the FDA will expect some testing to demonstrate that the new composition is still safe.

7.3.1.2 Food Additives

A food additive is "any substance the intended use of which results or may reasonably be expected to result – directly or indirectly – in its becoming a component or otherwise affecting the characteristics of any food" [4]. Any substance that is intended to be added to food or may come into contact with food is regulated as a food additive (direct or indirect). Direct food additives are those that are intended to have a specific purpose in food. Indirect food additives are not intended to be added to food but may become part of food in trace amounts. Food packaging, such as can coatings, plastic food containers, paper or paperboard packaging, and coatings for cookware, are all examples of indirect food additives.

New, direct food additives (including animal feed) require the submission and FDA approval of a food additive petition. Because food additives are intended to be consumed, the burden of proof for safety is significant. A "GRAS" (generally regarded as safe) determination, whether self-certified or submitted for FDA review, may seem like a shortcut to commercialization (compared to a food additive petition) because a GRAS determination can be done without FDA approval, but the data burden is essentially the same, and the safety assessment must be done by a panel of qualified experts. Furthermore, data used for the GRAS determination must be available to the public (e.g. published in a journal). Indirect food additives carry less of a burden, especially if the company can demonstrate that the substance does not migrate to food in any significant amount.

How long/how much? Innovators should expect to spend from $500 000–$2 000 000 and two to three years for testing and regulatory approval for a food additive petition or GRAS determination. Food contact notices range from $50 000–$500 000 and take one to two years. For food contact substances, migration calculations supporting an exemption can be done fairly quickly for minimal cost. If migration testing is necessary, a full suite of migration condition testing costs on the order of $10 000–$20 000 and should not take more than one to three months.

7.3.1.3 Cosmetics

Like other FDA-regulated products, the responsibility for safety of the product lies on the manufacturer of the final product (that is, the cosmetic product that is placed on the market). FFDCA defines cosmetics as products "intended to be rubbed, poured, sprinkled, or sprayed on, introduced into, or otherwise applied to the human body…for cleansing, beautifying, promoting attractiveness, or altering the appearance" [3]. Personal care products (shampoo, body wash) are considered cosmetics in the US. The law prohibits a manufacturer from marketing misbranded or adulterated products. FDA's website provides extended definitions of adulteration and misbranding [5], the key considerations of which are that the product must be safe and effective as intended and that labels and claims cannot be misleading.

The FDA does not impose a strict pre-market testing scheme for cosmetic ingredients other than certain hair dyes; nevertheless, the cosmetic industry has developed a suite of tests that the finished cosmetic manufacturers use to demonstrate safety. Cosmetic manufacturers generally expect ingredient suppliers to perform the testing necessary to document that the end product is safe. For many years, safety testing involved mammalian testing for skin and eye irritation. In 2019, testing on animals is still legal in the US, but most cosmetic manufacturers will not formulate products with ingredients that have been tested on animals. Instead, testing is done using a series of *in vitro* tests, often including skin and eye irritation, along with patch testing on human volunteers. This lack of a specific suite of tests that must be performed makes developing a pre-market test plan more challenging, but engaging with potential customers can provide needed clarity for the innovator.

The FDA has imposed restrictions on some specific cosmetic ingredients [6]. For example, chloroform and methylene chloride may not be used in any cosmetics (the regulations define any cosmetic containing either substance as an ingredient as adulterated). In other cases, the FDA restricts the use of some ingredients in particular products: Zirconium may not be used in any aerosol product due to concerns for inhalation toxicity.

How long/how much? Except for certain hair dyes, FDA does not perform pre-market review of cosmetic ingredients. The time and expense for introducing a new cosmetic ingredient into the market will depend on the requirements of the final cosmetic manufacturer. Testing may take 3–12 months and cost from $10 000–$100 000, depending on what data is available and what is needed to demonstrate safety. New manufacturing processes for existing cosmetic ingredients can probably be commercialized immediately if the innovator can demonstrate the safety of the new purity profile to the cosmetic manufacturer.

7.3.2 Federal Insecticide, Fungicide, and Rodenticide Act (FIFRA)

Pesticides are regulated in the US under FIFRA. Pesticides are defined as a "substance or mixture of substances intended for preventing, destroying, repelling, or mitigating any pest." Pesticides also includes products "intended for use as a plant regulator, defoliant, or desiccant" and nitrogen stabilizers (e.g. substances intended to stabilize nitrogenous fertilizer in soil) [7]. This definition includes both active pesticidal ingredients and the final formulations. States have overlapping jurisdiction with the EPA for pesticide regulations. Once a pesticide is registered with the EPA, manufacturers must also register the product for each state in which the product might be sold. For most states, state registration is more of a paperwork exercise than an additional testing or risk assessment burden, but it is an important pre-market step.

The manufacturer's intent is very important in determining whether a product is a pesticide. Considerations include claims on the product and in advertising material, including the product website, the composition of the product, the mode of action, and the end use. Pesticidal claims include indirect statements or implication of action against a pest or comparison to other pesticidal products – simply including a picture of a pest on a label may be considered a pesticidal claim.

Active ingredients and formulated end-use products must be registered with the EPA prior to placing products on the market. A manufacturer may commercialize a minimal risk pesticide without registering the product with the EPA if the pesticide meets the criteria set out in the regulations, but state registration may still be required [8].

Unlike products regulated by the FDA, intermediates to make pesticides are regulated by TSCA. In addition, if a pesticidal product is formulated in the US, all inert ingredients used in the formulation are also regulated by TSCA. By contrast, if a final formulation is imported into the US, TSCA does not have oversight, because the imported product (the final pesticide) is entirely under FIFRA jurisdiction.

How long/how much? Pesticides face significant regulatory scrutiny and require a substantial testing data set to support a registration. New conventional chemical active ingredients require the greatest amount of data, and innovators should plan for costs from

$500\,000–\$5\,000\,000$ and four to eight years for testing and regulatory approval, depending on many factors, including the end use. New inert ingredients must also be approved by the EPA, but the burden is usually comparatively less – on the order of six months to two years and $50\,000–\$200\,000$ of testing and fees. A new manufacturing process for pesticide ingredients may require notice to the EPA if the purity profile of the substance is different from that already approved, but the EPA may be able to approve the change without additional toxicity testing.

7.3.3 Toxic Substances Control Act (TSCA)

TSCA is the "catchall" control statute for chemical substances. Section 3 of TSCA states that "the term 'chemical substance' means any organic or inorganic substance of a particular molecular identity, including (i) any combination of such substances occurring in whole or in part as a result of a chemical reaction or occurring in nature, and (ii) any element or uncombined radical."[1] Basically, any atom or molecule in any oxidation state – a definition that a chemist would recognize. The definition then goes on to exclude the following:

- Mixtures (although components of mixtures are not excluded)
- Pesticides (as defined in FIFRA) when manufactured, processed, or distributed in commerce for use as a pesticide
- Tobacco or any tobacco product
- Any nuclear source material, special nuclear material, or byproduct material regulated by the Atomic Energy Act of 1954
- Firearms and ammunition that are regulated as such under the Internal Revenue Code
- Products regulated by FFDCA, including food, food additive, drug, cosmetic, or medical device when manufactured, processed, or distributed in commerce for use as a food, food additive, drug, cosmetic, or device

This definition is cleverly constructed in such a way that a product that is regulated under another authority remains under that authority, while products that are not regulated under another authority are regulated under TSCA. As stated in the introduction, if a product is used for more than one category of use, the product must meet the obligations for each of the intended uses. Ethanol, for example, can be an industrial solvent regulated by TSCA, a food, a drug, a pesticide active ingredient, a pesticide inert, and a cosmetic ingredient. Regulatory approval of ethanol for one of those uses does not allow ethanol to be used for any of the other uses. Innovators are often surprised to learn that a food ingredient that is being introduced for a TSCA use must go through TSCA review, even though it has been demonstrated as safe for food use.

TSCA divides chemical substances into two categories: existing substances and new substances. Existing substances are those that are listed on the TSCA Inventory, a list of about 85 000 substances that have been in commerce at some point since 1977. The EPA publishes a copy of the TSCA Inventory on its website [9]. Existing chemicals are subdivided into those that are active in commerce and those that are inactive. Inactive substances are those that have not been manufactured, imported, or processed in the US after June 21, 2006. Existing

---

1 US Code 15 § 2602.

substances that are active may be manufactured or imported for a nonexempt commercial purpose.

An entity that manufactures (a term that also includes importing) a substance is responsible for ensuring compliance with TSCA. Most substances listed on the TSCA Inventory do not carry specific restrictions and may be manufactured or imported by any entity without notifying the EPA. Because some of the substances listed on the TSCA Inventory have confidential substance identities, the EPA has a process for manufacturers to describe a substance and request that the EPA search for the substance on the confidential portion of the TSCA Inventory. The manufacturer must have a bona fide intent to manufacture the substance and demonstrate its intent in a Bona Fide Intent (BFI) to Manufacture notice. The EPA reviews the BFI to verify the substance identity, searches the TSCA Inventory, and, if found, responds with details that are necessary for the manufacturer to document compliance going forward.

If a substance is not listed on the TSCA Inventory, that substance is considered a new substance. The manufacturer must submit a new chemical notification (e.g. a premanufacture notice [PMN]) to the EPA and receive approval prior to manufacturing that substance in (or import to) the US for a nonexempt commercial purpose [10].

A PMN does not require that a submitter generate any test data, such as mammalian toxicity or toxicity to aquatic species, but any test data in the possession of the submitter must be included in the notice. Even though neither toxicity nor exposure information is required, it is critical that a submission provide the EPA with sufficient information that it can evaluate the potential risk to health or the environment. Such information may be from analogs (for hazard data) and from standard industry practices (for exposure information). This is especially true if the substance fits one of the New Chemical Categories [11]. The New Chemical Categories are groups of related substances for which the EPA has identified specific health or ecotoxicity concerns. Innovators must recognize the importance of submitting a robust PMN, especially for a substance that fits within one of the categories. If sufficient information is not included, the EPA's (justifiably) conservative estimates of toxicity and exposure can lead to significant delays in approval as the submitter seeks to fill data gaps (in hazard or exposure) after submission to demonstrate that there will not be unreasonable risk. After its review of a PMN, the EPA may require the submitter to develop specific data on toxicity or exposure to ensure that its predictions of toxicity or exposure were sufficiently protective, but such testing will be focused on the EPA's specific concerns, rather than a broad set of tests. In such cases, the EPA typically also imposes some specific restrictions on the conditions of use that led to the EPA's conclusion of unreasonable risk. These restrictions will be memorialized in a consent order (a negotiated legal agreement between the submitter and the EPA), a significant new use rule (SNUR), or both.

There are some exemptions to the requirement to submit a PMN [12]. Here is a partial list of exemptions:

- **Naturally occurring substances**. "Naturally occurring" has a narrow definition. To be considered naturally occurring, a substance must occur in nature and be unprocessed or processed "only by manual, mechanical, or gravitational means; by dissolution in water; by flotation; or by heating solely to remove water; is extracted from air by any means."[2]

---

2 40 C.F.R. §710.4(b).

If a substance meets the definition, it is automatically included on the TSCA Inventory and considered an existing substance.

- **Impurities** (substances that are unintentionally present in another substance).
- Certain **byproducts**.
- **Mixtures**. A mixture is exempt, but the individual component substances must be listed on the Inventory or otherwise exempt. The EPA considers hydrates of solid forms of substances as mixtures of the anhydrous form and water.[3] The EPA also has guidance on "statutory mixtures" – a class of products that, like hydrates, are formed by a chemical reaction but are nevertheless considered mixtures of substances [13]. Innovators are urged to seek expert assistance before relying on a statutory mixture exemption other than hydrates.
- **Articles**. A new chemical substance that is imported as part of an article is exempt from the PMN requirements. If a substance is intended to be released from an article, the substance would not meet the article exemption.
- **Nonisolated intermediates**. If an intermediate is produced and never removed from the equipment in which it was manufactured (and associated ancillary equipment such as pipes and pumps), and never held for storage, the substance may be eligible for the nonisolated intermediate exemption.
- **Export only**. A company may manufacture a new chemical substance if the entire amount of the substance is exported from the US prior to further reaction or use. Specific labeling and recordkeeping requirements apply.
- **R&D**. A company may manufacture small quantities of a new chemical substance for R&D as long as the R&D requirements are met [14]. The EPA's guidance does not specify what a "small quantity" is. Rather, a "small quantity" is relative to what a commercial quantity of the substance might be. More important in determining whether the quantity is appropriate is the consideration of the R&D activity itself. R&D can include performance testing of the substance, manufacturing at a pilot scale, or even trial runs at full scale. The larger the quantity produced, the more important it is to maintain documentation supporting that manufacturing at that scale supports R&D and is not merely commercial production.
- **Polymers** that meet the TSCA polymer exemption criteria [15] may be manufactured without an approved PMN. The polymer exemption criteria provide a basis to conclude that a particular polymer presents low hazard to health and the environment. Exempt polymers may be manufactured or imported at any time. The manufacturer must submit a note to the EPA the January following the first manufacture of an exempt polymer (or polymers).
- **Low Volume Exemption** (LVE) [16]. LVE requires notification and approval by the EPA prior to commercialization. The submission has similar information requirements as a PMN but is limited to uses specified and 10 metric tons or less manufactured per year. Exemption notices permit limited opportunities to change the parameters described in the notice. For example, an LVE manufacturer is bound by the specified use, manufacturing site, and release and exposure conditions. A manufacturer may request a "modification" of an LVE from the EPA to allow variance from the original submission, but it is

---

3 This definition excludes other reactions with water, such as hydrolysis products or reactions in which water is added across a carbon-carbon double bond.

an additions submission and an additional review fee. Exemption notices cost less than a third of the PMN fee and are reviewed in a fraction of the time of a PMN. Submitters are cautioned against submitting a PMN shorty after submitting an LVE. An LVE should be used only if the submitter genuinely expects the production volume not to exceed 10 metric tons in the first year and the EPA generally expects a manufacturer to operate under an LVE for at least a year before submitting a PMN. Each importer or manufacturer must have its own exemption. Substances manufactured or imported under an exemption are not added to the TSCA Inventory.

Substances listed on the TSCA Inventory that do not include a manufacturing process in the identity of the substance (the identity includes both the Chemical Abstracts Index Name and definition, if there is one) can be manufactured by any process without review by the EPA. Innovators need to pay particular attention to UVCB identities. Such identities may include a source (e.g. petroleum, or a particular plant species for natural oils), a process (e.g. hydrogenation), or both. Under TSCA nomenclature policy, if a substance includes source and/or process in the name, altering either the source or process leads to a different substance identity even if the final product is chemically indistinguishable.

How long/how much? From the time of submission of a PMN, a submitter should expect 6–12 months before being able to commercialize a substance. Some PMNs will be approved more promptly, especially substances that present low hazard to health and the environment. If the EPA finds that a substance is a potential risk for health or the environment, it will take some time for the EPA to take the necessary regulatory action (e.g., publish a SNUR) before allowing commercialization. A PMN has a filing fee of $16 000; exemption notices cost $4700. There are reduced fees for small businesses. While there are no up-front testing costs, the EPA may require testing after reviewing a PMN. TSCA requires that the EPA reduce vertebrate testing and also requires that the EPA take a tiered approach to testing.

7.3.3.1 Regulatory Gray Areas

Some products fall into regulatory gray areas – areas that challenge definitions and potentially challenge a "common sense" understanding of risks associated with end uses. A few examples are provided next. If you have questions about statutory jurisdiction, get help! A regulatory consultant can provide invaluable guidance as well as assist with preparation and submission of any necessary notification. The regulatory agencies may be able to assist, but a particular agency is likely to only answer questions related to its own jurisdiction.

- **Soap**: Ingredients in fatty acid salt-based soaps are regulated under TSCA, while ingredients in liquid hand soap and body wash are regulated as cosmetics under FFDCA.
- **Diapers**: Adult incontinence products are medical devices and are regulated under FFDCA; baby diapers are not medical devices and are regulated under TSCA. To be used in both products, a super-absorbent polymer must comply with both FFDCA and TSCA.

7.4 European Union (EU)

In the EU, chemicals that will be introduced for industrial or cosmetic uses are subject to Registration, Evaluation, Authorisation and Restriction of Chemicals (REACH) legislation.

Unlike TSCA in the US, REACH requires all manufacturers or importers of a substance to register that substance and share in the cost of testing. Furthermore, cosmetic ingredients must be registered under REACH as well as being approved for cosmetic use.

7.4.1   Registration

Substances that will be manufactured or imported in the EU above 1 ton per year must be registered. For each substance, one company assumes the role of lead registrant (LR). The LR is responsible for developing the substance identity profile (SIP), including a boundary composition for the substance. The LR also develops and submits a dossier for the substance. The dossier includes the details of the substance identity, the physicochemical properties, and toxicity and fate information. Higher tonnage volumes carry higher testing burdens. An LR can use a variety of strategies to satisfy the data requirements: testing, read-across (relying on data for substances that are sufficiently similar to the substance being registered), waivers (if a particular endpoint is not applicable to the substance), and quantitative structure-activity relationship (QSAR) models. A key aspect of REACH is that a data owner must be compensated if the owner's data are used to support a REACH registration. Once the registration is established, other manufacturers or importers may join the registration as long as the substance fits within the SIP. Joint registrants share the cost, whether it is sharing the cost of testing performed by the LR or sharing the cost of data access if the LR had to pay another data owner for access to a read-across substance.

A company can register at one of these four annex (tonnage) levels:

- Annex VII: 1–10 ton per year
- Annex VIII: 10–100 ton per year
- Annex IX: 100–1000 ton per year
- Annex X: More than 1000 ton per year

The manufacturer or importer must meet the registration obligations prior to exceeding the lower tonnage limit for the applicable tonnage band. For example, an importer may import up to 1 ton per year but cannot exceed 1 ton until the importer has registered the substance under Annex VII. Once registered at Annex VII, the importer may import up to 10 ton per year but may not exceed 10 ton per year until an Annex VIII registration is complete.

Annex VII registration requires a base set of physicochemical property testing, an acute oral mammalian toxicity study (another route may be performed if justified), an acute toxicity study in an aquatic invertebrate (typically daphnia), growth inhibition in aquatic plants (typically algae), *in vitro* gene mutation in bacteria (an Ames study), skin sensitization, *in vitro* skin and eye irritation, and a ready biodegradation study (if applicable to the substance). Each higher tonnage band adds more and higher-tier testing, adding significantly to the expense and time required to achieve registration. A full list of data points needed at each tonnage band is available on the European Chemicals Agency (ECHA) website [17].

Above 10 ton per year, a chemical safety report (CSR) is also required. A CSR includes details of the conditions of use for the substance, including worker exposures, releases to the environment, and potential general population exposures. The exposure levels are compared to the toxicity data to demonstrate that the conditions of use do not pose a risk.

The registration process begins with an inquiry dossier. A potential registrant might first search existing registrations for a substance that seems to match and attempt to demonstrate that the joint registrant's substance matches the SIP. The joint registrant builds an inquiry dossier with the necessary supporting information and submits that to the ECHA. The ECHA reviews the inquiry dossier and determines whether the substance does or does not match the SIP. If the substance does fit within the SIP, the joint registrant prepares a co-registration dossier by requesting and paying for access to the data. If the substance does not fit with an existing SIP, the registrant would begin the process of developing a lead registration and completing the applicable data requirements either by testing or by alternative mechanisms (e.g. read-across to substantially similar substance, data waiver, or modeling).

REACH has a radically different approach to polymers compared to other jurisdictions. In the EU, if a substance meets the definition of a polymer, the manufacturer or importer registers the monomer(s) (and other reactants, such as capping agents or neutralizing reagents) at the appropriate tonnage band (as opposed to the total polymer) [18]. The advantage of this approach is that innovators can "mix and match" monomers in myriad ways without seeking approval for each combination. For example, a polymer of two polyols A and B and three diacids, X, Y, and Z, would require registration of A, B, X, Y, and Z. Once the five substances are registered, dozens of polyesters can be manufactured with the various combinations. Although the REACH approach to monomers allows much more flexibility to manufacturers, the hazards associated with monomers are generally different than the hazards associated with polymers.

7.4.2 Exemptions to REACH Registration

The following categories are exempt from the REACH registration obligations:

- Food and feed
- Drugs
- Mixtures (although the individual components are subject to registration)
- Substances listed in Annex IV of REACH (minimum risk because of their basic properties, such as water, nitrogen, sucrose, and some monosaccharides)
- Substances listed in Annex V of REACH (including naturally occurring substances, most natural triglycerides, fatty acids, and their sodium, potassium, calcium, and magnesium salts)
- R&D (see next paragraph)

R&D. A company is exempt from the registration obligation for tonnages up to 1 ton per year. This tonnage is likely to satisfy most R&D needs. There is an additional exemption for companies that need to evaluate manufacturing at scale. The Product and Process Orientated Research and Development (PPORD) exemption allows larger tonnages for up to five years. Notification to the ECHA is required, but the testing obligations are not triggered. The ECHA evaluates the potential risks of the PPORD activity and grants the exemption from registration with any restrictions the ECHA deems necessary to protect health or the environment.

How long/how much? Obviously, the time and expense for registration strongly depend on how the registration is approached and the tonnage level. A joint registration can take as little as one to three months for the ECHA review of the inquiry dossier and to join the existing registration. A more complex substance may require more time to register, especially if extended discussion with the ECHA is necessary. Costs to join an existing registration depend on the total testing costs and number of other registrants and range from just a few thousand dollars to hundreds of thousands of dollars. An Annex VII (1–10 ton) lead registration requires on the order of 9 to 18 months and $100 000–$250 000. An Annex VIII lead registration is 18 to 24 months (can be concurrent with Annex VII) and costs $250 000–$500 000 in addition to the Annex VII testing. Higher tonnage levels become less predictable because whether or not higher-tier testing is required depends on the results of the lower-tier testing. For example, if the mutagenicity and screening-level reproductive and development toxicity testing are all negative, higher-tier reproductive toxicity testing will not be required. Furthermore, the ECHA will not permit the higher-tier reproductive toxicity testing to begin until the lower-tier testing is complete. Full Annex X testing can easily take three to five years and cost millions of dollars.

7.5 China

The Ministry of Environmental Protection Order No. 7, issued by China's Ministry of Ecology and Environment, is the primary law regulating the control of chemical substances placed on the Chinese market. China takes a hybrid approach to chemical control. Like the US, there is an inventory of substances that may be manufactured or imported in China, the Inventory of Existing Chemical Substances Produced or Imported in China (IECSC). If a substance is not listed on the IECSC, a manufacturer or importer must register the new substance in a process that is similar to REACH, with required testing at various tonnage tiers. As with REACH, cosmetic ingredients must be on the IECSC, or registered, in addition to being approved for use as cosmetics. China revised its environmental regulations as of January 1, 2021. Some of the requirements remain the same as those described here.

7.5.1 Registration

There are three types of notifications for new chemicals: scientific research record (SRR), simplified, and regular notifications [19]. Unlike REACH and TSCA, R&D substances are not exempt from pre-market notification. SRR notifications are for substances intended for R&D uses under 100 kg per year. Submitting a complete SRR notification is fairly straightforward, and once submitted, the substance may be manufactured or imported – no review by the Chinese authority is required.

Simplified notifications (special) (SNS) are available for polymers of low concern, substances used as intermediates or for export only up to 1 ton per year, R&D substances up to 1 ton per year, or process R&D up to 10 ton per year over two years. As with an SRR notification, testing is not required to support the SNS.

Simplified notifications (general) (SNG) are available for substances that are not eligible for SNS. SNG applies to substances produced up to 1 ton per year. Testing, including an acute fish or earthworm toxicity study and ready biodegradation study, are required to support an SNG.

Above 1 ton per year, regular notification is required. Regular notices and requirements for supporting data are similar to REACH registrations, with some notable differences:

- Full study reports must be submitted (that is, there is not a data sharing provision as in the EU).
- Skin and eye irritation must be *in vivo*.
- Aquatic toxicity testing to support Chinese registration must be done by labs in China.
- Substances subject to notification are typically added to the IECSC five years after notification.

A company can register at one of these four tonnage levels:

- Level 1: 1–10 ton per year
- Level 2: 10–100 ton per year
- Level 3: 100–1000 ton per year
- Level 4: Greater than 1000 ton per year

7.5.1.1 Exemptions
As with the US and the EU, China does have exemptions to the requirement to register new chemical substances. Notable exemptions include the following:

- Naturally occurring substances
- Substances subject to other laws and regulations (e.g. pesticides and drugs)
- Nonisolated intermediates

Polymers are not exempt from registration, although polymers that meet the criteria for low concern are eligible for SNS. The polymer of low concern criteria are similar to the TSCA polymer exemption criteria.

How long/how much? There is a wide range of timeframes and costs associated with registration in China. SNS are straightforward and can be completed fairly promptly, in as little as a week. Regular notifications, like REACH registrations, depend on the tonnage level, and the costs and timeframes are similar.

7.6 Canada

In Canada, industrial chemicals are regulated under the Canadian Environmental Protection Act (CEPA) of 1999. CEPA is like TSCA in many ways [20]. The most notable difference is that CEPA requires a specific set of testing at specific tonnage thresholds. Like TSCA, CEPA has an inventory, called the Domestic Substances List (DSL). The list of exemptions is similar as well. CEPA also has a "pre-inventory" called the Non-Domestic Substances List (NDSL) that is comprised of all the substances listed on the public portion of the TSCA Inventory that are not also listed on the DSL. Substances that are not on the DSL must be submitted to Environment and Climate Change Canada (ECCC) in a New Substance Notification (NSN) prior to exceeding the applicable tonnage threshold. Substances that are on the NDSL may be moved to the DSL with less of a regulatory burden (generally higher initial tonnage limits and lower testing requirements).

CEPA divides substances into two overall categories: chemicals (and biochemicals) and polymers (including biopolymers). Biochemicals, a subset of chemicals, are substances produced by a microorganism for a commercial purpose. Similarly, biopolymers are polymers produced by a microorganism for a commercial purpose [21]. Substances extracted from whole plants or animals (e.g. essential oils, cellulose) are not considered biochemicals or biopolymers [22].

Like the US and China, CEPA allows for a lower reporting burden for polymers of low concern. The criteria for Reduced Regulatory Requirement (RRR) polymers are similar to the TSCA polymer exemption criteria.

The reporting thresholds and timelines depend on the substance type and tonnage. Table 7.1 lays out the range of reporting obligations. Schedule 1, 3, 4, and 9 NSNs require only basic information about the substance – no testing is required. Schedule 5 requires basic physicochemical testing, an Ames test, acute aquatic toxicity, and acute mammalian toxicity testing. Schedule 6 has a significantly greater testing burden, adding aquatic

Table 7.1 NSN reporting obligations.

NSN Schedule	Category	Minimum days prior to exceeding tonnage threshold	Annual quantity (kg)
Chemicals			
1	Special category[a]	30	1 000
4	Not on NDSL	5	100
4	on NDSL	30	1 000
5	Not on NDSL	60	1 000
5	on NDSL	60	10 000
5	on NDSL – high release/exposure[b]	75	50 000
6	Not on NDSL	75	10 000
Polymers			
3	Special category[b]	30	10 000
9	All polymers	30	1 000
10	Non-RRR polymers on NDSL or all reactants on DLS or NDSL	60	10 000
10	Non-RRR polymers on NDSL or all reactants on DLS or NDSL – high release/high exposure[b]	60	50 000
11	Non-RRR polymers not on NDSL or not all reactants on DLS or NDSL	60	10 000

a) R&D, contained site-limited intermediates, and contained export-only substances.
b) Additional evaluation if >50 000 kg/y and either >3 kg/day release to water or significant public exposure is anticipated.
Source: Guidelines for the Notification and Testing of New Substances: Chemicals and Polymers. Government of Canada. © 2019 Government of Canada [22].

toxicity, genotoxicity, skin irritation and skin sensitization, acute mammalian via another route, and repeat dose mammalian toxicity testing. Testing under Schedule 10 (for polymers) is similar to Schedule 5 (less the Ames test). Testing under Schedule 11 is similar to Schedule 6 (less the additional acute mammalian study). As with the EU, read-across or waivers for specific testing may be justified.

How long/how much? Canadian NSNs are among the most predictable for new chemical review. The data set required is known in advance, and submission fees are quite modest (less than $4000). Testing to support a Schedule 5 or 10 is about $20 000–$50 000 and requires two to six months. Testing to support a Schedule 6 or 11 is on the order of $75 000–$200 000.

7.7 Developing a Global Strategy

Given the time necessary to complete testing and notification in each of these sample jurisdictions, innovators need to plan ahead to ensure that each regulation is satisfied prior to commercial launch in a given jurisdiction. Review the possible markets (by use and country), examine the pre-market burden for each, and develop a plan of how to maximize market penetration with minimal up-front costs. For example, if a product can be launched in a single market to develop an income stream, that income can pay for the testing to launch in additional markets.

Part of jurisdictional strategy work involves a careful data gap analysis.

- Make a list of all physicochemical and toxicity endpoints that will be needed to support registration or notification in each country and for each statute (e.g. TSCA, FFDCA, REACH).
- Review what endpoints may be waived and, if possible, develop the waiver justification for each regulatory authority.
- Review what data is available, either in the public domain or owned by others, that can potentially be used to support a read-across argument.
- For each remaining data point required, determine how long testing for each endpoint might take at the various tonnage levels and how quickly the market might grow over that timeframe.
- Be sure to consider whether particular jurisdictions or customers require, allow, or prohibit testing on vertebrates.

With a complete data gap analysis, a test plan can be implemented. The plan may have some duplicate testing (e.g., aquatic toxicity testing in China and other jurisdictions) but will generally give a clear picture of what testing must be performed to achieve pre-market approval and allow the most efficient path to market launch.

7.8 Summary

Global chemical control regulations can be complex and even counterintuitive, but compliance is critically important for innovators seeking to commercialize their chemical technologies around the world. Satisfying the pre-market requirements does not necessarily have to delay commercial launch, but a deep understanding of the regulatory requirements and a plan to do testing and seek approval from each authority is vital in ensuring a smooth path to market. By looking and planning ahead, all the regulatory requirements can be met in a timely and efficient manner, allowing initial market penetration to help pay for expansion into other markets or a greater market share.

Assistance with pre-market approvals is available from a variety of sources, including government agencies, regulatory consultants, and law firms. This assistance may be in the form of a turnkey service, where the consultant performs the majority of the work or a consultancy service helps the innovator develop the capacity in-house. In any event, novices to regulatory submissions are well-advised to seek assistance from experts. Mistakes made in the pre-market process can lead to delays in approval or factual errors in the registration or approval that have to be corrected, potentially disrupting the business. It is far better to get it right the first time.

References

1 Bergeson, L.L. (ed.) (2014). *Global Chemical Control Handbook: A Guide to Chemical Management Programs*. s.l.: American Bar Association. ISBN: 978-1-62722-739-1.
2 US Food and Drug Administration. (2019). General controls for medical devices. https://www.fda.gov/medical-devices/regulatory-controls/general-controls-medical-devices (accessed 30 October 2020).
3 US Food and Drug Administration. (2019). Is it a cosmetic, a drug, or both? (Or is it soap?). https://www.fda.gov/cosmetics/cosmetics-laws-regulations/it-cosmetic-drug-or-both-or-it-soap (accessed 30 October 2020).
4 US Food and Drug Administration. (2019). Overview of food ingredients, additives & colors. https://www.fda.gov/food/food-ingredients-packaging/overview-food-ingredients-additives-colors (accessed 30 October 2020).
5 US Food and Drug Administration. (2019). FDA authority over cosmetics: how cosmetics are not fda-approved, but are fda-regulated. https://www.fda.gov/cosmetics/cosmetics-laws-regulations/fda-authority-over-cosmetics-how-cosmetics-are-not-fda-approved-are-fda-regulated#What_kinds (accessed 30 October 2020).
6 US Government Printing Office. (2019). Code of Federal Regulations. Subpart B—Requirements for Specific Cosmetic Products. 21 C.F.R. §700.
7 US Environmental Protection Agency. (2019). What is a pesticide? https://www.epa.gov/minimum-risk-pesticides/what-pesticide (accessed 30 October 2020).
8 US Environmental Protection Agency. (2019). Conditions for minimum risk pesticides. https://www.epa.gov/minimum-risk-pesticides/conditions-minimum-risk-pesticides (accessed 30 October 2020).
9 US Environmental Protection Agency. (2019). TSCA chemical substance inventory. https://www.epa.gov/tsca-inventory (accessed 30 October 2020).
10 Engler, R.E. (2017). Procedures and Approaches for New Chemical and Significant New Use Notification, Review, and Regulation. In: *New TSCA: A Guide to the Lautenberg Chemical Safety Act and Its Implementation* (eds. L.L. Bergeson and C.M. Auer). Washington: American Bar Association.

11 US Environmental Protection Agency. (2014). TSCA New Chemicals Program (NCP) chemical categories. https://www.epa.gov/sites/production/files/2014-10/documents/ncp_chemical_categories_august_2010_version_0.pdf (accessed 30 October 2020).

12 Campbell, L.M. and Burchi, L.R. (2017). Chemical Substances Not Found on the TSCA Inventory but Exempt from PMN Requirements. In: *New TSCA: A Guide to the Lautenberg Chemical Safety Act and Its Implementation* (eds. L.L. Bergeson and C.M. Auer). Washington: American Bar Association.

13 US Environmental Protection Agency. (2019). Products containing two or more substances, formulated and statutory mixtures on the TSCA Inventory. https://www.epa.gov/tsca-inventory/products-containing-two-or-more-substances-formulated-and-statutory-mixtures-tsca (accessed 30 October 2020).

14 US Environmental Protection Agency. (1986). *New Chemicals Information Bulletin: Exemptions for Research and Development and Test Marketing*. Washington, DC: US Environmental Protection Agency.

15 US Environmental Protection Agency. (2019). Polymer exemption for new chemicals under the Toxic Substances Control Act (TSCA). https://www.epa.gov/reviewing-new-chemicals-under-toxic-substances-control-act-tsca/polymer-exemption-new-chemicals (accessed 30 October 2020).

16 US Environmental Protection Agency. (2019). Low volume exemption for new chemical review under TSCA. https://www.epa.gov/reviewing-new-chemicals-under-toxic-substances-control-act-tsca/low-volume-exemption-new-chemical (accessed 30 October 2020).

17 European Chemicals Agency. (2019). Information requirements. https://echa.europa.eu/regulations/reach/registration/information-requirements (accessed 30 October 2020).

18 European Chemicals Agency. (2019). Guidance for monomers and polymers. https://echa.europa.eu/documents/10162/23036412/polymers_en.pdf (accessed 30 October 2020).

19 MacDougall, L.S., Burgess, A.G., and Bergeson, L.L. (2014). *Global Chemcal Control Handbook* (ed. L.L. Bergeson). Washington: American Bar Association.

20 Burchi, L.R. (2014). *Global Chemical Control Handbook: A Guide to Chemical Management Programs* (ed. L.L. Bergeson). Washington: American Bar Association.

21 Government of Canada. (2005). New substances notification regulations (chemicals and polymers). (2005). *Justice Laws Website (Goverment of Canada)*. https://laws-lois.justice.gc.ca/eng/regulations/SOR-2005-247/FullText.html/ (accessed 30 October 2020).

22 Government of Canada. (2019). Guidelines for the notification and testing of new substances: chemicals and polymers. http://publications.gc.ca/collections/Collection/En84-25-2005E.pdf (accessed 30 October 2020).

Part III

Springing into Action

8

Navigating Supply Chains
Tess Fennelly

GreenSustains, Inc., Minneapolis, MN 55311, USA

8.1   Introduction

In Chapter 2, you did your homework on market segmentation, your customer needs, your competitors, and your value propositions and competencies. You developed a path for your business using your differentiated offering to potential customers. However, your business won't be operating in a vacuum. A critical success factor is defining and understanding your supply chains and channels: the field in which you and your competition will be operating.

There continues to be growing interest and awareness in sustainable chemistry technologies, particularly the adoption of less hazardous alternatives to existing products. Demands from both regulators and customers will continue to fuel the demand for these safer products. However, despite efforts from many stakeholders to accelerate their deployment, adoption rates of these technologies remain relatively low. Commercialization in the chemical space is inherently slow, often measured in years. The innovator needs to be aware that some of this delay can be attributed to supply chain challenges, which are often overlooked. In this chapter, we will discuss how to identify key hurdles in the supply chain and ways to overcome them.

8.2   Supply Chain Complexity

The complexity of supply chains in general is enormous. This is compounded even further when considering technologies that are based on alternative chemistries, formulations, compounds, and fabrication methods. In addition, end-use applications and markets for these products are diverse, including textiles, electronics, packaging, cleaning products, paints and coatings, adhesives, automotive, and many more. Each have their own established supply chains that contain thousands of chemicals and processes that convert raw materials into literally millions of products. To complicate things further, each position along this supply chain has many suppliers globally. Figure 8.1 gives a single example of a supply chain and its complexity, in this case for a fiber textile product.

How to Commercialize Chemical Technologies for a Sustainable Future, First Edition.
Edited by Timothy J. Clark and Andrew S. Pasternak.
© 2021 John Wiley & Sons Ltd. Published 2021 by John Wiley & Sons Ltd.

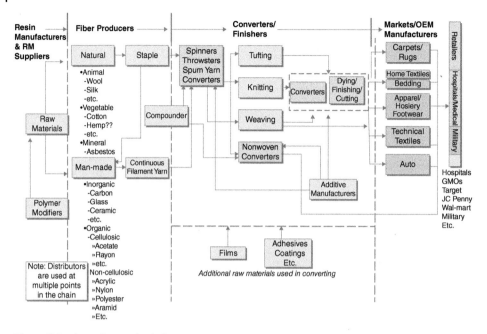

Figure 8.1 A textile supply chain.

As you can see, the various flows in this supply chain are certainly complex. Individual suppliers exist in each phase; some serve multiple positions, while others serve individual positions.

In Chapter 2, we examined customized market segmentation and groups of customers. Within any particular supply chain, potential customers could be either decision-makers or influencers. Figure 8.2 breaks down the example textile supply chain by identifying general decision-makers (green boxes) and influencers (yellow boxes) for an additive manufacturer (black box).

Now take this a step further. Imagine you are producing a new additive that targets the Carpets/Rugs segment. To gain a better understanding of the supply chain, the key first step is to map out the elements for the Carpets/Rugs box. Some can be readily determined. For example, you know from your business dealings that the fibers used to make carpets and rugs are dominated by nylon, polyester, and wool. Other routes may be more difficult to discern depending on the details of your technology. For example, fiber producers' opinions could be impactful depending on how the additive affects the uptake or retention of dye in the fiber. Perhaps the compounders find the additive causes an undesirable migration of other additives; in such a case, compounders would be strong influencers. If the final carpeting is used in residential, commercial, automotive, or aircraft applications, the additive may have regulatory restrictions (e.g. the additive should not smoke or generate toxins when burned). There could be further issues with performance at different temperatures, use within existing fabrication equipment, or with its ability to be recycled. All of these factors have potential implications in the supply chain.

Once you have a rough map of the supply chain , you can then conduct your customer needs interviews (described in Chapter 2) to help fully develop a greater understanding

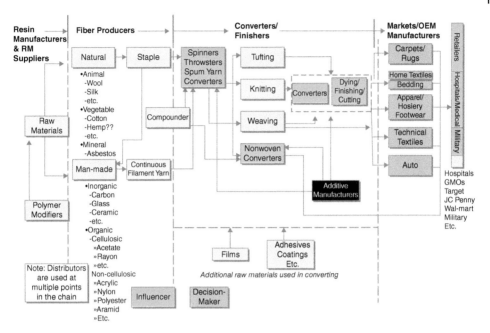

Figure 8.2 Decision-makers and influencers for additive manufacturers in the textile supply chain.

of each supply chain. Place the key companies that supply products/services under each box in the chain and ensure that you include your competition as well. This is a "living process" – as you learn more, your supply chain map will be fine-tuned. You can then strategically decide which specific customers to serve, design product offerings that best meets their needs, and identify and exploit advantages over your competition.

Once you are ready to sell your product, the greater understanding gained from this process will enable you to make a positive impression on the supply chain influencers and, in turn, the decision-makers. It's also pretty difficult to figure out where to go and who to work with if you don't have a clear path on where and how to operate.

As you become more knowledgeable about the market segment and the flow of products/services, this supply chain map can become a valuable tool by including pricing, costs, and margin received at each key position within the chain.

8.3   Recognizing Points of View

We've established that supply chains are complex, and understanding them is critical to navigate your market segments and meet your customer needs. What is often overlooked is that each position in the supply chain provides a different viewpoint for companies. Each company (and even individuals within a given company) in your supply chain often view identical things with different perspectives due to past experience, personal bias, and a plethora of other subjective factors.

To illustrate the "point of view" concept graphically, what do you see when looking at Figure 8.3?

Figure 8.3   What do you see?

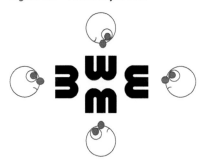

Figure 8.4   What do you see now?

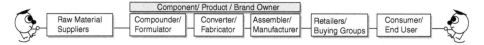

Figure 8.5   Supply chain positioning produces different points of view.

Is it a "m"? It could be, but it depends on your point of view. Now look at Figure 8.4; this is the same graphic as Figure 8.3, but depending on the viewpoint, the graphic could be a "m," "w," "3," or "E."

Understanding points of view is particularly important in the supply chain for sustainable chemical technologies. In the simplified supply chain in Figure 8.5, each player has a different point of view regarding this field.

For example, if asked "how important is sustainable chemistry to your business?" many chemical companies will say "extremely." This is a natural response; why wouldn't it be important? But this general approach risks causing confusion and misinterpretation of your customers' needs. You are assuming each position even defines sustainable chemistry in a clear and consistent way. There is a good chance this is not true. Before beginning any conversation with customers, clearly define what you both mean when discussing sustainable technologies.

Thus, the first question you ask should be "what does sustainable chemistry mean to you, and which elements of that definition are most important to your business?" It is likely the answers you receive will differ greatly. A converter/fabricator may talk about conforming to retailer and regulatory consumer product requirements, EPA Safer Choice, transparency in listing ingredients, and impact on the recycling stream. However, a raw material supplier might discuss life-cycle analysis, biorenewable content, and raw material availability. No perspective is wrong; it is just their differing realities based on their subjective experiences and needs. These different points of view are at least in part informed by what they see and hear from their customers, suppliers, and influencers.

Misunderstanding points of view is one of the most common pitfalls in marketing and sales. It is easy to be enamored with your technology and have tunnel vision on your market offering. However, yours is a singular point of view. At every step of the process, closely listen to the feedback you receive from others in the chain to ensure you stay grounded and well-informed on what is important to them.

> If there is any one secret of success, it lies in the ability to get the other person's point of view and see things from that person's angle as well as your own
>
> -Henry Ford

8.4 Supply Chain Hurdles and Strategies to Overcome Them

Beyond the sheer complexity of any individual supply chain, there are a wide number of other challenges that likely need to be addressed, as shown in Figure 8.6.

These hurdles apply pressure and break the flow and uptake of new sustainable chemical technologies in the chain. Identifying these barriers to entry for your new technology and ways to navigate and overcome them will play a key role in the success of your business. Each of these will be addressed individually in the following sections.

8.4.1 Incumbency: Incumbents and Legacy Suppliers Own the Supply Chain, Market Access, and Global Supply

8.4.1.1 The Challenge

No matter which supply chain you enter, there are always existing players with established operating assets. These could be locally or globally positioned within established distribution networks. They undoubtedly have established raw material streams and optimized production processes with multiple manufacturing sources for the same product. They also have validated process economics, well-understood product performance data, market expertise, global regulatory registrations, and established customers. In short, they have an enormous wealth of credibility in the market. Additionally, some may also capitalize on their country's natural resources, low labor costs, and government subsidies or ownership.

Customers know their suppliers and the supply chain. Strong working relationships will have been established at multiple points in the chain including, in some cases, the raw material supplier. Legacy suppliers often provide a wide portfolio of complementary products and have established shipping, distribution, and seamless supply processes in place. Customers are comfortable with this established infrastructure and its predictability, consequently resisting change. This poses a challenge for all new market entrants to compete within the supply chain.

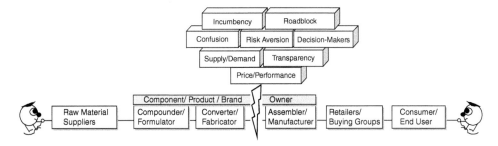

Figure 8.6 Deterrents to the adoption of sustainable chemical technologies.

8.4.1.2 How to Overcome It

One way to address this challenge is to simultaneously engage as many stakeholders in the supply chain as possible. It is important not to get trapped in the "go-alone" mindset. If your value proposition is well-defined, incumbents may be recruited to be allies in your battle for market acceptance. It will also serve to unify industry acceptance and help accelerate adoption of your technology. Another strategy is to study the incumbents and enter the market in a nonstrategic portion of their target market. Getting a foothold early without alerting or upsetting the giants will go a long way in building market presence and eventual growth.

8.4.2 Roadblock: Adoption Must Occur at all Points in the Supply Chain in Order to Be Successful

8.4.2.1 The Challenge

Broad adoption of a technology occurs only if all points in the chain are on board. No matter how amazing your product is and the value it offers, any gaps in the supply chain will be a strong roadblock to its adoption.

Global brands have the market influence and sheer scale to rapidly mobilize new sustainable chemical technologies. The zeal and excitement are tangible on both sides when global brands partner with early-stage companies rooted in sustainable chemistry (Figure 8.7). Partnering with a global brand is generally a huge success for the entrepreneur, as discussed in greater detail in Chapter 9. However, even this is often not enough.

For example, several beverage and snack companies have partnered with biopolymer start-ups to commercialize bio-based compostable and biodegradable packaging. This generated great anticipation for rapid adoption of biopolymer technologies. However, supply chain resistance slowed adoption.

There were technical challenges that downstream processors needed to address such as changes to throughput due to slower processing times and changes to processing temperatures to accommodate the new polymer's processing window.

The general skills and expertise required to compound and process these biopolymers was limited. Capital was needed to make changes to the manufacturing assets to process these materials. The compounders and fabricators pushed back due to these issues, which created a roadblock that significantly slowed the commercialization of these polymers (Figure 8.8).

While successes may have been announced by the material and brand owners, the delay and pushback from the middle of the chain have been challenges to overcome.

8.4.2.2 How to Overcome It

The key lesson here is that full supply chain partnership and alignment are needed to address the reduction in production costs and to create a strategy for the required investment. Without full supply chain alliances, the path toward commercialization of

Figure 8.7 Global brands partnering with smaller sustainable chemistry companies is on the rise.

Figure 8.8   New technology uptake roadblocks.

new sustainable chemical technologies will continue to be difficult. This includes the small innovator who must be willing to partner with the suppliers, processors, and end users within the chain. Make sure you have mapped the supply chain as described earlier so you can identify and speak with the key players operating between your technology and the ultimate customer.

8.4.3   Confusion: "Green Washing," Perceptions, and Misinformation Confuses the Industry and Consumers on What Is Truly More Sustainable, Which Impacts Demand

8.4.3.1   The Challenge
Roadblocks can also be caused by misinformation and "green washing." For example, although the term "toxicity" produces a universal reaction of "bad," many conflicting campaigns are producing widespread confusion about what specific materials fall into that category. Another example is the "chemicals are bad" mentality and the rise of "chemo-phobia." This has caused an aversion to, or prejudice against, chemicals or chemistry and created uncertainty and confusion that have hampered the adoption of sustainable chemical technologies.

An example of this is a sustainable chemistry start-up that invented a new class of bio-based solvents. The solvents have such a safe profile that they could be used in the personal care industry for skin contact. They had no hazardous volatile organic compounds (VOCs), no odor and posed little human health toxicity. The solvents had powerful solubilizing capabilities, dissolving some of the most difficult crystalline ingredients and had excellent non-drying-skin emollience. Additionally, they were soluble in water and oil making them a versatile personal care formulation tool.

Excited to enter the market, the start-up purchased booth space at the Society of Cosmetic Sciences – Supplier Day. Supplier Day is one of the most important North American events, with more than 300 global cosmetics and personal-care industry companies showcasing their new innovative ingredients and solutions to the decision-makers at personal-care product customers. Booth banners and product literature were prepared showing case data and applications of this new product. Unfortunately, the literature contained the word "solvent," and when customers saw this, they immediately thought of hazardous and toxic properties. This start-up learned that renaming their ingredients as solubilizers or emollients was the key to selling into this segment. The product was in fact a safe solvent, but the mindset of "solvents are bad" was too entrenched to overcome.

Conflicting information and opinions also continue to confuse the broader industry, as illustrated by the example shown in Figure 8.9.

This confusion can cause supply chain stakeholders to be grid-locked and hesitant to implement changes due to uncertainty about the status quo of current products.

Figure 8.9 Formaldehyde. American Chemical Council says "safe" [1]. American Cancer Society says "not safe" [2]. Source: American Cancer Society.

Even when the use of toxic chemicals is restricted owing to regulatory status, regrettable substitutes appear with labeling that confuses customers. Figure 8.10 depicts a new paint stripper that quickly arrived on shelves after the US Environmental Protection Agency (EPA) announced a retail ban of "methylene chloride" from all retail paint stripping products. The "non methylene chloride formula" text front and center on the new formulation could lead to consumers assuming that this new formulation is less hazardous. However, it is hazardous. The product is flammable, it causes burns and eye damage, its inhalation may cause cancer, and prolonged exposure may lead to organ damage.

This conflicting information and how it is marketed confuses the industry and consumers on what is truly more sustainable. This confusion can impact the growth of, and demand for, new and more sustainable products.

8.4.3.2 How to Overcome It

The best defense is to study the landscape, anticipate competitive responses to your technology, and work with commercialization partners that are dedicated to sustainable alternatives. Partnering with a credible industry player currently serving these channels is another way to accelerate entrance and market acceptance. Furthermore, you should have integrity and back up your sustainability claims with solid data – be it technical, LCA, or other recognized means as detailed in Chapter 3. You can't control how the competition markets their product, but you can control yours so differentiate based on best practices.

8.4.4 Risk Aversion: Worries of Failure Due to Poor Performance, Brand Tarnishing, Hidden Costs, and Stagnant Product Sales

8.4.4.1 The Challenge
There is often concern that switching to an alternative, more sustainable chemical technology could lead to market failure for a host of potential reasons. These include technical risks such as poor performance, hidden costs (process or equipment changes), and material incompatibility. Nontechnical risks may be even greater and include brand tarnishing in the event of product failure, additional required workforce training, and customer education. There is also the risk of adopting a "better bad," as depicted in the paint stripper example in Figure 8.10. Real or not, these are all potential risks that any new adopter will likely consider.

8.4.4.2 How to Overcome It
Stakeholders are cautious and will naturally ensure that they have the information needed to make sound new material management decisions. This conservative approach slows the growth and adoption of new sustainable chemical technologies. The best defense is doing your due diligence to really understand customers' concerns and points of view. Creative contract elements such as a lead time exclusivity, first adopter discount, or partnership can help to speed uptake. Once one customer in a segment has successfully adopted the technology, the risk is mitigated for those following. Getting the first deal is key, even if it isn't the best deal. Finding a regional player or a midsize supplier can be an excellent way to enter the market, learn, and then grow.

Figure 8.10 Substitute paint stripper in response to methylene chloride ban. Source: http://www.kleanstrip.com/product/ks-non-methylene-chloride-premium-stripper.

8.4.5 Decision-Makers: Sustainable Corporate Objectives vs. Operations May Not Align

8.4.5.1 The Challenge

Sustainability has been on the agenda at many large corporations for decades. Despite best intentions at the highest levels, there are still mismatches among the decision-makers in these larger companies. Many now have "sustainability directors" who are responsible for identifying initiatives to improve the sustainability performance of the company. Unfortunately, the goals and incentives for these sustainability directors often do not align with those of individuals responsible for budgeting and operating performance (e.g. ensuring the business generates profits). For them, the risk of not achieving growth in the business unit category is simply too large to take. This causes companies to struggle to implement sustainability into the operational profit centers of their business. This is especially true in times of corporate tension. Breaking this barrier and driving changes in purchasing practices is a key hinderance to adoption of sustainable chemical technologies in the supply chain.

Furthermore, senior executives often find that changing a supply chain to use more sustainable materials is likely to come at a price with at least some short-term investments. The result of the added up-front costs (even with long term pay-out) means that there can be added resistance to making the change [3, 4].

8.4.5.2 How to Overcome It

The good news is that the sustainability movement has created an opening for dialogue and the introduction of new sustainable chemical technologies at corporate levels. It has also given rise to global protocols to assist sustainability practices. One example of this is the UN Sustainable Procurement Statement that targets the incorporation of sustainable purchasing practices (Figure 8.11) as a way to accelerate adoption.

Actions are being taken by more progressive corporations to implement responsible material procurement. These now require their supply chain leaders to employ sustainable purchasing practices. This in turn will impact how procurement departments are evaluated at corporate levels. In addition to their performance metrics and targets based on cost reduction, metrics that consider the savings achieved through sustainability will also be considered beyond the unit price. For example, changing material specifications to enable better recycling yields may not change the unit price but will still result in savings based on its total sustainable impact.

UNITED NATIONS
GLOBAL MARKETPLACE SUSTAINABLE PROCUREMENT

- Procurement is called sustainable when it integrates requirements, specifications and criteria that are compatible and in favor of the protection of the environment, of social progress and in support of economic development, namely by seeking resource efficiency, improving the quality of products and services and ultimately optimizing costs.
- Through sustainable procurement, organizations use their own buying power to give a signal to the market in favor of sustainability and base their choice of goods and services on:
  - economic considerations: best value for money, price, quality, availability, functionality;
  - environmental considerations, i.e. green procurement: the impacts on the environment that the product and/or service has over its whole life-cycle, from cradle to grave;
  - and social considerations: effects of purchasing decisions on issues such as poverty eradication, international equity in the distribution of resources, labour conditions, human rights.
- Sustainable procurement is not about "burdening" the market with extra requirements; rather it is a well-defined strategy that gradually phases in sustainable requirements in tenders and bids, promotes dialogue and open communication between the suppliers and procurers.

Figure 8.11 UN sustainable procurement statement. Source: UN Sustainable Procurement Statement. © 2020 United Nations.

This provides an opportunity for the innovator. Targeting corporations that incorporate these metrics into their purchasing practices as your first customers will help your sustainable chemical technology gain traction in the market.

8.4.6 Supply and Demand: Concern in Committing to a Single Sourced New Technology

8.4.6.1 The Challenge

A key barrier to displacement of materials of concern is the lack of supply of safer sustainable alternatives. While many start-ups and established firms have developed such alternatives, most of these have not had a significant impact in displacing major segments of the market as they are the sole supplier of these materials. Larger companies will not risk their product lines unless they can be guaranteed a steady supply of material at a constant price.

Furthermore, sustainable chemical technology supply and demand is a "chicken and egg" dilemma. New material manufacturing investments take years and may cost hundreds of millions of dollars. Material suppliers often find there is not enough real or perceived demand to justify this investment. Conversely, producers and customers expect a new technology to ramp quickly, perform seamlessly, be available from more than one source, and meet large volume needs. Uncertain supply and demand commitments can jeopardize timing and investment in the supply, thus slowing the availability and adoption of new sustainable chemical technologies. These lower volumes mean poor economies of scale that inhibit cost competitiveness with incumbent solutions.

Processors and producers are key to building demand but are typically reluctant due to concerns related to sole sourcing and limited or lack of global supply. Incurring transition and requalification costs for a source that may not get traction or grow is risky. Once the new product is approved or qualified, the fabricators and customers are often impatient at the slow growth of supply infrastructure.

8.4.6.2 How to Overcome It

To help mitigate some of these challenges, it is important to create strong partnerships and make sure that all expectations are clearly understood between both parties. This is particularly true of first customers. Try to establish a mutual long-term planning process and agree to long-term commitments that make sense for both sides. These can ensure that supply infrastructure growth will be smoother and faster. For a meaningful partnership, both parties must see and work toward a win/win scenario.

8.4.7 Transparency: How to Satisfy Customer and Regulatory Demands While Protecting Intellectual Property and Trade Secrets

8.4.7.1 The Challenge

Transparency requires the full disclosure of all chemical components/formulants in a finished product. It may seem reasonable to provide this high level of visibility to customers. However, transparency demands can stifle innovation and lead to conflicts within the supply chain.

For example, sustainable material suppliers struggle with balancing the requested transparency from their customers versus the need to protect their intellectual property (IP,

> Cleaning Product Right to Know Act' becomes law in California
>
> Manufacturers of cleaning products sold in California will be required by law to disclose the existence of certain chemicals in their products, making the state the first in the nation to pass such legislation.
> The law requires ingredients to be shared on manufacturer websites by Jan. 1, 2020, and on labels one year later. Companies can protect trade secrets, but only if the chemicals they want hidden have not been linked to harmful effects on humans and the environment.

Figure 8.12 California's Right to Know Act as an example of transparency.

patents, or trade secrets), as described in Chapter 4. The innovator's decision to release this information can result in the following:

- Loss to low-cost imports
- Technology knock-offs
- Competitors filing blanket intellectual property (IP) around the innovator's IP, tying up commercialization
- Loss from costly legal proceedings with larger corporations to defend the innovator's IP position and technology
- Loss of competitive position when trade secret formulations are made public

This need for transparency has fueled growth in stakeholder and regulatory bodies that have the power to really affect the innovator's commercialization timelines. For example, local or national authorities may issue mandates based on restricted substances lists or create incentives for purchasing and product placement. A recent example is the Right to Know legislation passed in California (Figure 8.12).

The need for openness will continue to increase, and blocking transparency is not a good option as it can lead to concerns and potential liability for the unknown composition. It can also slow industry acceptance in segments requiring transparency. In some cases, transparency can prevent commercialization since a customer may need to formulate their product for national distribution while requiring compliance with individual state laws. This "us versus them" struggle has impacted the flow and adoption of new technologies.

8.4.7.2 How to Overcome It

The best defense for new material suppliers is to make sure that the technology is truly ready for introduction. Some early testing can be done under Confidential Disclosure Agreements and Material Transfer Agreements. Early commercialization can occur with some of these customers under contract, particularly in industrial applications. This will allow the new technology to gain some early traction and for you to learn what else is needed before broad commercialization.

However, pay close attention to state or regional rules similar to the Right to Know legislation in California. In certain applications, legislation will not allow these products to be commercialized without disclosure. For example, California's Right to Know disclosure requires the following:

- A list of each intentionally added ingredient contained in the product with CAS number (if applicable) in order of predominance down to 1% (below 1% the order does not matter).
- A list of all "nonfunctional constituents" (there are 34 of them listed in the Act) present in the designated product at a concentration at or above 0.01% (100 ppm).
- The functional purpose served by each intentionally added ingredient.
- A list of all fragrance ingredients that are included on a designated list, including allergens (at or above 0.01% [100 ppm]).
- Electronic links for designated lists will be grouped together in a single location for any intentionally added ingredient or nonfunctional constituent that is included on a designated list.
- A link to the safety data sheet for the designated product.

8.4.8 Price/Performance: It's More Than Price per Pound; Total Cost Savings Need to Be Communicated

8.4.8.1 The Challenge

Although large retailers and brand owners are pushing for sustainable chemical technologies, there continues to be limited willingness to accept any change in price or performance. This is far and away the most cited reason for the slow adoption of new technologies.

A significant degradation in performance or increase in cost is going to be unacceptable. However, there may be some room for compromise on price and performance. For example, there are often savings in a total cost analysis such as reduced hazardous waste handling and disposal and a positive impact on the corporate sustainability scorecard. Unfortunately, this can be hard to quantify for customers focused only on cost per pound pricing.

8.4.8.2 How to Overcome It

Approaching new customers as collaborative partners is the best way to mitigate the pricing discussion. First, understand where there are long term savings for the customer. Will they be able to avoid expensive personal protective equipment (PPE) for the line workers or no longer need to contract for hazardous waste disposal or air scrubbers? Will they be meeting an unmet need with their customers? Work together with your customer to calculate these savings. This total cost analysis will work toward making your technology more attractive and help your customer to have an even stronger story for promoting the technology internally.

Working with customers on the first-generation product and iterating with them to target continuous improvement is often seen as a plausible way to enter the market even with higher-priced materials. This is even more acceptable if you have already calculated the total cost analysis with your customer. The segments for first introduction may be smaller, but their valuable feedback can be used to tweak the technology to meet their needs in terms of performance and price. An improvement over the status quo may not necessarily represent the ultimate goal but certainly provides a step in the right direction.

One of the classic examples of this is Henry Ford's famous statement regarding his Model T automobile:

> People can have the Model T in any color – so long as it's black.

Although it may have worked at the time, this strategy obviously wasn't feasible in the long run. Continuous improvement and product generations later, Ford vehicles come in every color and design imaginable.

8.5 Lessons Learned

Take the time to map and understand the complexity of your supply chains. Identify the needs of each segment based on both its customers' and suppliers' points of view. Understand the many potential hurdles that may affect each segment and develop strategies to overcome them as described here. It may be time-consuming, hard work, but it is actually one of the fastest ways to create opportunities to accelerate the commercialization of your sustainable chemical technology. Your successful entry to the market will be driven by how you respond to the dynamics of the supply chain, your ability to navigate its hurdles, your willingness to collaborate and your adeptness in creating opportunities.

References

1 American Chemistry Council. (n.d.). The polycarbonate/BPA global group. https://www.americanchemistry.com/Product-Groups-and-Stats/PolycarbonateBPA-Global-Group (accessed 30 October 2020).
2 Zissu, A. (2016). 9 ways to avoid hormone-disrupting chemicals. NRDC. https://www.nrdc.org/stories/9-ways-avoid-hormone-disrupting-chemicals (accessed 30 October 2020).
3 Frazee, G. (2019). 4 reasons it's hard to become a sustainable business. PBS NewsHour. https://www.pbs.org/newshour/economy/making-sense/4-reasons-its-hard-to-become-a-sustainable-business (accessed 30 October 2020).
4 Winston, A. (2018). Explaining the business case for sustainability again … and again … and again. https://sloanreview.mit.edu/article/explaining-the-business-case-for-sustainability-again-and-again-and-again (accessed 30 October 2020).

9

Strategic Partnering

Jason Clark[1] and Shawn Jones[2]

[1] Braskem America, Cambridge, MA, USA
[2] White Dog Labs Inc., New Castle, DE, USA

9.1 Introduction

9.1.1 Partnerships as a Change Driver

Relationships, within the context of business activity, come in many different forms. Certain relationships are legal in nature, such as establishing a limited liability partnership (LLP). Others are financial, such as a start-up and its investors, while still others are transactional, such as the relationship between a customer and a service provider. Another category, known as "strategic partnerships," seeks to employ aspects from each party toward a mutually beneficial goal such as expanding a market, introducing a new product, or defending against a competitive solution.

Deloitte identified 35 social, technological, environmental, economic, and political drivers of change at the macroscale that are expected to shape the future [1]. Included in this list, the *growth of partnerships was specifically identified as an important change driver* that itself may result from other change drivers. For example, consider the technological advances in the areas of artificial intelligence, automation, and digitalization more broadly. Each of these technologies has application and impact in the chemical industry but depends on skill sets that have not traditionally been a focus area for chemical manufacturers. Strategic partnerships involving start-ups, academia, and other established companies have facilitated the adoption of digital technologies by traditional chemical companies, many of which now have dedicated teams.

Additionally, change drivers that include increased competition for talent, crowdsourcing, the growth of the knowledge worker, and the maturation of the next-generation workforce all reinforce the value of strategic partnerships. The roles and expectations within the employee-employer relationship have undergone a dramatic shift since the baby-boomer generation, forcing many companies to evaluate how they attract and retain the best talent or otherwise gain access to talent in a competitive way [2]. Strategic partnerships with other entities is one of the tools that can help companies be competitive in a challenging labor market, access specialized skill sets, and take advantage of alternative perspectives to their challenges.

How to Commercialize Chemical Technologies for a Sustainable Future, First Edition.
Edited by Timothy J. Clark and Andrew S. Pasternak.
© 2021 John Wiley & Sons Ltd. Published 2021 by John Wiley & Sons Ltd.

Finally, the change drivers related to increasing environmental awareness, growing concern over climate change, an increased regulatory environment, and resource scarcity directly underpin the need for new sustainable chemical technologies. These particular challenges are large in scope, affecting the entire globe and many aspects of daily life. As no single company is capable of tackling these challenges alone, strategic partnerships are likely a prerequisite to making significant progress. The International Council of Chemical Associations' report on the efforts of the global chemical industry toward the UN's Sustainable Development Goals (SDGs) describes the many ways that partnerships are working to develop sustainable chemical technologies [3].

9.1.2 Partnerships for Sustainable Chemical Technologies

It is clear that sustainable chemical technologies are a priority for many industrial companies. The goals are specific to the company, but they encompass many different aspects of the business and the communities in which the companies operate. The following statements are extracted from sustainability reports from ExxonMobil, BASF, and DuPont, respectively:

> ExxonMobil's primary responsibility is to produce the energy and products the world needs in a responsible manner. Our approach to sustainability focuses on six key areas: Corporate governance, safety, health and the workplace, managing the risks of climate change, environmental performance, community engagement and human rights, local development and supply chain management. [4]

> Chemistry for a sustainable future: Business success tomorrow means creating value for the environment, society and business. Our innovations contribute to a sustainable future. We support our customers in being more sustainable through our solutions and create new business opportunities that reinforce our customer relationships and attract new customers. In this way, we also contribute to achieving the U.N. Sustainable Development Goals (SDGs), which were adopted by the United Nations as globally recognized economic, environmental and social objectives. [5]

> Our sustainability framework has three strategic focus areas: Innovation to thrive, create sustainable solutions to society's most pressing challenges; Sustainable Operations, deliver world-class, end-to-end performance in safety, resource efficiency and environmental protection; People and Well-being, enable the health and well-being of people and communities and advance diversity and inclusion in our workforce and beyond. [6]

Achieving these ambitious sustainability goals will certainly rely heavily on innovation. A recent survey of 270 corporate leaders identified 10 key challenges that large companies encounter when trying to innovate [7].

- Politics, turf wars, and a lack of internal alignment
- Cultural issues
- Inability to catch critical signals or developments
- Inability to act on critical signals or developments
- Lack of budget

- Lack of strategy or vision
- Not adopting emerging technologies
- Lack of executive support
- Recruiting
- Lack of CEO support

There is an opportunity for strategic partnerships to exploit potential advantages byaddressing the sustainability goals of each company and the broader challenges regarding innovation.

9.1.3 Chapter Structure

This chapter will address a number of key issues related to strategic partnering. First, the potential partnership advantages/disadvantages and mitigation of risks will be presented from the perspective of the start-up and the industrial partner. Next, the evaluation, establishment, execution, and closure of strategic partnerships will be described. Finally, a series of case studies will be examined to demonstrate the lessons of the chapter with practical examples.

9.2 Advantages and Disadvantages of Strategic Partnering

Engaging in a partnership involves a trade-off between potential benefits of working closely with an external partner versus its disadvantages and the cost of continuing independently. The following table summarizes the pros and cons from the perspective of the start-up and the larger partnering company. Each of these will be discussed in greater detail in the following sections.

	Potential Advantages	Potential Disadvantages
For the Start-Up	• Providing additional talent • Access to scale-up resources • Market access • Development guidance • Industrial validation	• Loss of control over development • Limiting additional investors • Reduced market reactivity
For the Industrial Partner	• Building an innovation culture • Access to new capabilities • Easier implementation of new business opportunities	• Intellectual property challenges • Control balancing challenges • Organizational disruption

9.3 The Start-Up Perspective: Partnership Advantages and Disadvantages

9.3.1 Partnership Advantages for the Start-Up

For a start-up to grow, capital is required to fund continued development, hire more talent, acquire new equipment, or scale up. Initial capital often comes from angel investors, venture funds, or government grants. A strategic partnership is an additional option once the start-up has demonstrated its product or process works (at least at the lab-scale) and identified potential partners their product or process would impact. While the partnership may bring needed capital, there are several additional advantages a strategic partnership can bring a start-up over other funding sources.

9.3.1.1 Advantage 1: Providing Additional Talent

By entering a strategic partnership, the industrial partner is almost always committing some of their own personnel to develop the technology. At a minimum, the partner may commit scientists or engineers to review and discuss results of the project and provide guidance to the start-up. At the other end of the spectrum, the partner will embed personnel within the start-up to work full-time to advance the project. For the start-up, not only does this defray additional personnel costs, but also relieves the need to recruit and hire new personnel, which can be difficult for a lean start-up to achieve.

In addition to technical personnel, a partner may also offer specialized support personnel, such as regulatory experts, logistical specialists, or intellectual property specialists. These skillsets are often beyond the scope of a start-up's hiring needs and often fall to consultants. Similar to a new full-time hire, start-ups must first find the necessary consultant, vet their credentials and background, negotiate a contract, and pay for the offered services. By accessing the partner's internal experts on these matters, the start-up can save significant time and capital.

9.3.1.2 Advantage 2: Access to Scale-Up Resources

One of the most difficult transitions a start-up has to make is from lab-scale to pilot or demonstration-scale. This evolution requires new equipment, new personnel (or retraining of existing personnel), and often reworks to the process. Most start-ups simply lack the resources to build or acquire their own pilot facilities, so one option is to contract an external toll manufacturer or other scale-up facility. While this relieves the need to purchase their own equipment, it does leave the start-up reliant on the existing equipment and timelines of others.

A strategic partnership offers the start-up another option to achieve pilot or demonstration-scale without these disadvantages. Unlike the start-up, the strategic partner can use its existing facilities for piloting or demonstrating the necessary process. Additionally, they have the flexibility to augment or update the equipment for the start-up's exact process – something a traditional toll manufacturer may not be willing or able to do. Finally, the start-up can interact directly with the operators of the facility to gain knowledge of how the system operates and performs. This more collaborative relationship can better address any new operational challenges. The start-up may still need to work around existing time and resource demands of the industrial partner, but if there is enough internal drive by the strategic partner, it is far more likely to adjust timing. In an ideal situation, the industrial partner would agree to build or retrofit a dedicated facility for the joint partnership. In this case, the start-up can potentially gain knowledge not only on how the process is run but also on how the facility is designed, built, and operated.

9.3.1.3 Advantage 3: Market Access

Rarely do start-ups develop a completely novel process independent of the current manufacturing landscape. Rather, the new product or process must be integrated into the existing infrastructure. This can present a significant challenge. Even if a direct replacement product has been made, the start-up must still convince a customer to switch their current process or supplier for a new one. This challenge is even greater for a novel process or product. Additionally, a start-up may not fully understand the interconnectivity of various processes and suppliers, which makes this task even more difficult. Not only can this interconnectedness involve financial and logistical relationships but also personal relationships that need to be cultivated. The technology may work exactly as advertised, but if a start-up cannot gain traction within the marketplace, it will ultimately fail.

This "selling" of the innovation must take place. The question is do you (i) make a pitch to a strategic partner that your technology is worth their investment, or (ii) sell the technology to the industry partner once it is mature and validated. An advantage of the former is that it can be pursued before a significant amount of internal resource investment. If a partnership cannot be achieved, the start-up can still pursue development using its own resources (e.g. venture funding or federal grants) and possibly try again once a later milestone is reached. Once a partnership is agreed, both the start-up and industrial partner can pursue a marketplace adoption in parallel while the process is being developed. The industrial partner can bring other stakeholders into the conversation (e.g. purchasing department, subsidiary manufacturers, other industrial partners). While all new technologies still face risks in the market, both the start-up and industrial partner can maximize the chances of success by working together to deploy the technology of interest.

9.3.1.4 Advantage 4: Development Guidance

Start-ups are often founded and driven by an entrepreneur with a vision for their technology. While this is an exceptional benefit, it can also blind the entrepreneur to how their technology will actually fit into the industrial environment. For example, does the technology require a small change in current manufacturing practices? The entrepreneur may respond with "Yes, but it's a small change, and the benefits far outweigh any costs incurred." In reality, that change can have a ripple effect upon the entire process rendering the scheme unworkable. These conclusions by the entrepreneur are often made due to a lack of knowledge and can have a devastating impact on the success of the start-up.

One of the most meaningful impacts a strategic partnership can have for a start-up is the industrial guidance it can provide. Is the technology being developed actually needed by the industry? Does it solve an actual problem or is the start-up creating a problem just for it to solve? Are the assumptions being made by the start-up correct? Though the answers to these questions may be difficult for an entrepreneur and their team to hear, they are critical for the success of the start-up and better to be addressed sooner rather than later. If the start-up is willing to listen and address the critiques of the industrial partner, they can improve their technology faster and ensure market relevance. This is not to say the industrial partner cannot be wrong or make mistakes too. Rather, if the entrepreneur or their team disagrees with a recommendation or conclusion by the industrial partner, they should make their own case back to the partner or check with other consultants in the space. This iterative cycle has the potential to improve not only the developed technology but both the start-up's and strategic partner's teams as they are forced to validate their own assumptions.

9.3.1.5   Advantage 5: Industrial Validation

By entering into a strategic partnership with a large and (ideally) credible company, the start-up and its technology immediately gain a greater amount of credibility. Potential investors and customers now view the start-up very differently. Instead of toiling alone in obscurity, they are working with a real partner with broad recognition in the market. People will naturally see the formation of the partnership as a measure of validity. Why would the partner decide to invest resources into an unknown start-up if there wasn't potential value in their technology? This additional credibility goes a long way in opening new doors for the start-up and can be used as an effective means of entering into new relationships.

9.3.2   Partnership Disadvantages for the Start-Up

9.3.2.1   Disadvantage 1: Loss of Control Over Development

As the industrial partner brings more and more resources to bear on the project (e.g. technical experts, regulatory specialists, pilot or demonstration facilities), they can begin to view the project more as an internal R&D project rather than a partnership with the start-up. From the start-up's perspective, this may be perceived as a narrowing of the scope of the technology. Rather than a broad, industry-changing application, the strategic partner may want a more narrow, targeted application suitable for their specific interests. This can come about from the industry's relatively conservative, long-term outlook that if the narrow application is successful, further adoption in other application spaces can be pursued. Additionally, the fast-paced environment of a start-up can be slowed by the conservative nature of the strategic partner. Whereas two or three successful runs may have previously been acceptable to the start-up to justify the next stage of development, the industrial partner may require several more trial runs to ensure repeatability. This perspective can leave the start-up feeling restricted and that the strategic partner is limiting what they can achieve.

Another concern is the potentially slow decision-making process of the larger strategic partner. While the immediate managers and engineers on the project may work at the start-up's pace, the corporate office may not. This ultimately impacts timelines. Slow decisions on go/no-go points, funding, use of facilities or equipment, and personnel allocation can again feel restricting to the entrepreneur and lead to resentment.

9.3.2.2   Disadvantage 2: Limiting Additional Investors

In addition to the technical restrictions a strategic partner may impose on a start-up, they can also have an indirect effect on potential investors. By signing with one industrial partner, the start-up is limiting its ability to work with other strategics in a similar space and, potentially, in other spaces. For example, it would be difficult to secure a second strategic partnership with a rival commercial player even in a different field because the original commercial partner could feel challenged by this action. Thus, once the strategic partnership is signed, the start-up is largely committing to that one player. Similarly, venture funds may not be interested in investing depending on the arrangements of the partnership. As highlighted earlier, the industrial partner could have a potential negative effect upon the pace or scope of work, which makes the start-up less attractive to venture investors. Additionally, if the strategic partner keeps drawing out the deployment of the technology, the payback for the venture fund also gets drawn out, which can lead to resentment on the part of the investor.

9.3.2.3 Disadvantage 3: Reduced Market Reactivity

Start-up teams are almost constantly innovating. Even while one technology is being developed, entrepreneurs are already thinking of another technology or application to pursue. One of the most potentially impactful disadvantages of a strategic partnership is the loss of flexibility to pursue these interests. This can be particularly true for an innovation or discovery that is beyond the scope of the strategic partnership. If the start-up were not in a strategic partnership, it could present the potential opportunity to its board, and a swift decision can be made to either stay the course or pursue the new innovation. However, once a strategic partnership is signed, flexibility is reduced. While the industrial partner may understand the importance of the discovery, it may be beyond their market scope. They may simply not be in a position to share the potential benefits due to this market mismatch. If this occurs, the partner would still insist the start-up meet its agreed upon commitments to the partnership and not pursue the innovation, at least with the resources it provides. It may be possible for the start-up to access different funds to pursue the innovation, but can be challenging within a strategic partnership.

9.4 The Industrial Partner Perspective: Partnership Advantages and Disadvantages

9.4.1 Partnership Advantages for the Industrial Partner

9.4.1.1 Advantage 1: Building an Innovation Culture

Operational excellence is often cited as a key success factor for industrial companies [8]. Focusing on operational excellence results in systems, processes, and procedures that ensure safety. It also minimizes business risks and increases predictability. However, unfortunate consequences of focusing on operational excellence are often the creation of barriers to innovation. These barriers can subsequently hamper the implementation of innovative products or processes the company needs to grow. The creation of barriers is especially true when the innovation may cannibalize parts of the core business. They can be further exacerbated by leadership revisiting memories of unsuccessful prior efforts when approached with a new idea. Forming a strategic partnership offers an opportunity to access innovation not subject to the same barriers, as well as incubate a culture shift that may spread to the larger organization over time.

An additional benefit to the flexibility built into the partnership is the employees of the larger organization are exposed to a culture that has high agility, adaptability, and tolerance to failure. By experiencing this culture and operating in a different environment, employees can gain an understanding of the benefits of both cultures and learn when to apply different aspects to future challenges. The expanded view of these employees can help minimize the impact of past unsuccessful efforts and can energize people, continually pushing the company toward a vision of an improved innovation culture [9].

9.4.1.2 Advantage 2: Access to New Capabilities

Accessing new capabilities is a key advantage to establishing partnerships with start-ups as they can significantly help an industrial partner take action, especially in new fields experiencing rapid development.

The partnership between BASF and Citrine Informatics to apply artificial intelligence in the chemical industry illustrates this advantage [10]. BASF identified areas relevant to its business where digitalization will have an impact [11]. One of these, accelerated environmental catalyst development, was selected for execution in a partnership with Citrine. BASF will contribute experimental data and expertise while Citrine will supply the machine learning competence and capability. In the absence of a partnership, BASF might have (i) established the capability internally, (ii) utilized a service provider, or (iii) not acted at all. Developing internally represents a significant financial commitment that includes the on-boarding of new employees with the appropriate skillsets. Thus, the speed of implementation would be tied to bringing these new individuals into the company, which is not a small task.

Furthermore, managing the consequences of an unsuccessful project on the people and assets can be challenging. The use of a service provider may mitigate the resource challenges but may also require extended timelines if the service does not yet exist. Having to wait until the service is available creates a risk that a competitor who has partnered to co-develop the solution is able to take advantage of the technology sooner. By approaching this challenge via partnership, BASF's research and development programs are able to immediately benefit while also limiting their risk to the financial and expertise contributions.

In addition, the industrial company will also develop a strategic roadmap for securing the needed capabilities from the partnership. The roadmap may include a combination of options such as continuing with the partnership, pursuing an acquisition, making focused knowledge transfer efforts, or pursuing internal development. The choice is scenario-specific but can be tailored to the risk level, resource availability, and time requirements of the industrial company.

9.4.1.3 Advantage 3: Easier Implementation of New Business Opportunities

Lack of action by a large company on an opportunity may also be caused by lack of leadership support or clear alignment with current business objectives. Consider the following example related to annual objectives for a chemical plant manager:

- Achieve site-wide safety performance of zero recordable injuries and zero process safety incidents throughout the fiscal year.
- Achieve asset utilization rate of at least 95%, excluding the two-week maintenance turn-around.
- Complete the maintenance turnaround with less than 10% budget over-run and no time over-run.
- Achieve a reduction in operational fixed costs of $100 per ton by year-end in line with the expectation of achieving first quartile cost performance in three years.
- Achieve 90% compliance of site employees with diversity and inclusion training requirements by end of the second quarter.

While these objectives may comply with the S.M.A.R.T (specific, measurable, achievable, relevant, time-oriented) philosophy, they are likely to be a deterrent for the adoption of a new sustainable chemical technology. The risks inherent with any new chemical technology may be a threat to these objectives, while potential benefits are not captured to offset the risks. A company may certainly adjust their approach to objectives and restructure itself for increased flexibility, but these changes are not quick, easy, or without their own risks [12]. Strategic partnerships offer an additional route to action that can begin in the near term and adjust as the industrial company evolves.

For example, a new brand can be established as part of the partnership. This new brand provides a distinct value proposition for the novel sustainable chemical technology that is separate from the industrial companies existing offerings. The industrial company can therefore avoid complicating their existing brand's value proposition while also allowing for separate, perhaps unconventional, pricing models. Additionally, if the partnership is established discreetly, the industrial company may also utilize the partnership to explore markets of strategic interest without revealing this interest to potential competitors. Finally, the partnership can be used to evaluate new business models.

The value proposition of sustainable chemical products has additional dimensions compared to standard chemical products. The use of alternative sales methods and marketing strategies may be advantageous. An example of using a strategic partnership for new business is the joint development between the industrial company DIC and the start-up Checkerspot for the development of polyols [13]. DIC benefits from access to new high-performance materials and from Checkerspot's ability to evaluate a market response through a newly developed premium ski product [14], something that DIC is unlikely to attempt within its normal business operations.

9.4.2 Partnership Disadvantages for the Industrial Partner

As with any business decision there are trade-offs and forming strategic partnerships are no exception. The advantages highlighted previously make a strong case for the pursuit of these relationships, but they must also be weighed against the potential disadvantages.

9.4.2.1 Disadvantage 1: Intellectual Property Challenges

Intellectual property (IP) is a key consideration for any business. As described in Chapter 4, IP may take many different forms including copyrights, patents, trademarks, and trade secrets. IP creates differentiation in markets and barriers to competition. It is also frequently subject to challenge, and the consequences for violation of IP rights can be severe [15].

When entering into a strategic partnership, a series of IP considerations may arise. The first is the handling and exchange of confidential information. This frequently occurs both to and from the industrial partner and is governed by a nondisclosure agreement (NDA) or confidential-disclosure agreement (CDA). The exchange of confidential information represents two additional risks: (i) information from the industrial company is inappropriately disclosed to third parties and (ii) information disclosed to the industrial company contaminates other areas of their work. To address these risks, the company may need to maintain a separate team that interfaces with the start-up. It may also be necessary to cease internal developments on related IP to prevent conflicts or alternatively to incorporate these existing efforts into the partnership. In either case, the industrial partner must practice good information management and constantly balance the benefit of disclosing or receiving confidential information related to the joint development against the risk of third-party disclosure or contamination [16].

Ownership and access of IP developed during the partnership may also present challenges. The terms of ownership must be addressed during the establishment phase of the partnership and can take many forms. Both parties are encouraged to seek the advice of experienced IP counsel when forming the partnership.

9.4.2.2 Disadvantage 2: Control Challenges

A partnership brings together and aligns the capabilities and competencies of each entity toward a common goal. Details of the governance, objectives, deliverables, and duration establish the operating framework for the partnership. Balancing the level of control that each entity has over these different aspects greatly influences the degree of satisfaction with the partnership outcomes.

Consider first a scenario where the industrial partner has a high level of control. In this scenario, it is likely that the partnership will adopt a governance that resembles that of the industrial partner, perhaps to the extent where the start-up functions more like a contract service provider. This may negatively impact the ability of the start-up to provide the aforementioned benefits such as speed, adaptability, and creativity as they now face similar challenges of the teams within the industrial partner. This may be compounded by geographical differences and communication practices. Furthermore, the industrial partner may limit the exploration of opportunities outside of the original scope. For an early-stage start-up that is seeking market fit for its sustainable chemical technology, this may impede their maturation process.

In the opposite scenario, where the industrial partner exhibits too little control in the partnership, negative outcomes such as drift in scope and slippage in timeline or budget may occur. The industrial partner may have unique requirements that are unknown to the startup and introducing them during the project will inevitably affect the timeline or budget. Similarly, without the benefit of the industrial partner's expertise to anticipate certain risks and facilitate mitigation plans, a reactive approach is employed which will also likely impact both timeline and budget. Resource allocations within a start-up are highly dynamic and includes activities outside of the partnership. The industrial partner must proactively engage to ensure prioritization of the partnership related efforts in the face of other pressures. This includes the pressure to modify the scope of the partnership activities to allow the start-up to address other needs or opportunities that ultimately deliver results not aligned with the industrial partner's expectations.

9.4.2.3 Disadvantage 3: Organizational Disruption

In a study evaluating open innovation at large organizations, the management of organizational change was identified as the primary challenge of engaging in open innovation [17]. Both the establishment and management of a partnership necessarily involve many different activities of the industrial company and have different requirements than an internal effort. Resource allocation within the industrial company and performance management may also experience disruption to the normal processes.

For example, challenges may arise when the required resources are not under the direction of the area seeking the partnership. Consider a partnership where the industrial company is interested in using the sustainable chemical technology from a start-up in an application that complements the industrial company's core products – perhaps a new polymer additive or a new surfactant for laundry detergent. The product or business development areas are the likely source for this partnership, but the start-up may require the support of the industrial company's process engineering expertise to achieve commercial-scale operation. The reallocation of these resources from internal efforts to developing technology within the partnership introduces a new dimension for resource prioritization for which leadership may lack guidance. The business strategy must be adapted to include both the pursuit of partnerships and the constraints in which the partnerships may operate. The constraints themselves may be dynamic and through regular review can be adapted and communicated through the management chain to facilitate proper resource prioritization.

Continuing with the earlier scenario, if a process engineer from a manufacturing site within the industrial company is now dedicated to the partnership via on-site relocation, the individual performance objectives must change. An individual's change in performance objectives is likely manageable within existing systems. The challenges introduced by the partnership centers on who is the proper person to evaluate the process engineer's performance and how will the evaluation be handled. Clearly, the manufacturing site management is no longer appropriate to evaluate the process engineer's performance. However, the product or business development team that originated the partnership may not have the expertise to evaluate their performance. In this scenario, the industrial company must also have systems in place for managing remote employees that also touches on areas of benefits, compensation, and compliance. These are examples of organizational disruptions stemming from individual participation in the partnership; these disruptions may also be experienced in the performance management of company systems.

9.5   Mitigation of the Disadvantages and Risks

With the most prominent disadvantages and risks for both parties understood, clearly defined strategies to mitigate them will now be discussed.

9.5.1   For the Start-Up

Most of the disadvantages highlighted essentially involve the entrepreneur feeling like they have lost some measure of control. This could be related to the overall company vision, use of resources, or decision flexibility. Therefore, it is critical before entering into a strategic partnership that the start-up not only has a clear understanding of the goals and timelines of the collaboration but also of the culture of the industrial partner. If it is known that decision-making can be slow, deadlines should be set to provide sufficient time between decisions such that work does not need to stop. Likewise, funding periods and allocations should be set to ensure resources are not limiting the work. In addition, the entrepreneur and their team should be open to learning from the strategic partner and using their critiques to further improve the technology.

9.5.2   For the Industrial Partner

Adopting certain practices can help mitigate the impact of the aforementioned disadvantages. The first recommended practice is to identify an executive-level sponsor for the overall strategy of engaging in partnerships. Executive sponsorship can help ensure that the many areas of an industrial company that may be affected by strategic partnerships understand the value and are aligned toward adopting any organizational changes necessary to be effective in these partnerships.

The second recommended practice is to identify a senior leader for sponsorship of individual strategic partnerships. Senior leadership engagement can help ensure that the partnership objectives are aligned with the industrial company's strategy, helping to streamline the negotiation and approval process. Additionally, it can help define constraints on the scope and highlight areas of freedom. Senior leadership can also motivate active participation of employees in the partnership, facilitating the dynamic balance of control and identifying additional ways of creating value from the partnership.

The third recommended practice is to step completely through the business model for delivering value from the partnership during the establishment phase. The process of thinking through both the immediate and longer-term needs will help identify what may be expected of each partner as the partnership progresses. Key negotiation points such as the management of intellectual property, the type and quantity of resources to be made available, and the commercialization strategy should be clarified. While moving to negotiations without adopting this practice may result in executing the partnership agreement faster, there is an increased risk that unmanaged terms arise during the partnership that require renegotiation. Subsequent negotiations are likely to introduce delays or perhaps end the partnership due to a lack of agreeable terms without achieving the desired objectives and after the expenditure of resources.

The final recommendation, although its implementation was highlighted as a potential disadvantage, is to adopt specific key performance indicators (KPIs) and metrics for strategic partnerships. While a champion-driven strategic partnership program can obtain positive results, it may not capture the longer-term benefits associated with internal cultural improvement and increased innovation that can be monitored with a systems approach. Gaining insight into the impacts beyond the partnership objectives will help the industrial company to actively manage its portfolio of partnerships such that the efforts will continually improve and adapt with the overall company strategy.

9.6 Evaluating a Potential Partnership

9.6.1 Start-Up Perspective

When and how to pursue a partnership is an important decision for a start-up. There are no specific metrics for when or how to pursue a strategic partnership, but there are several questions to consider before committing to one.

First, how broad is your technology? If it has potential applications across multiple industries, finding one strategic partner that shares your vision can be difficult. Venture funding may be more open to a broadly disruptive technology but doesn't bring the same type of capabilities and competencies. If your technology is very focused on one specific industry or process, finding a strategic partner could make more sense. In a similar vein, how competitive is the marketplace? If it is dominated by one player, are they known for innovation and inventiveness? If not, pursuing a partnership with them may be counterproductive, but courting a smaller competitor may make sense to give them an advantage.

Second, is the technology at the appropriate stage to pursue a partnership? A start-up needs at least validation data demonstrating their technology works, but beyond that there is no specific criteria before seeking a partnership. You should understand the market well enough to make a credible case for how your technology is unique, the advantages and disadvantages it offers, and a plan to get to market. This last point is important as it sets the entrepreneur apart from the academic investigator. The academic investigator can invent a truly groundbreaking sustainable chemical technology, but if they have no vision of how to implement it within the manufacturing landscape, it remains just a unique technology that will never deliver on its potential sustainability benefits. The entrepreneur can provide the justification and steps necessary to actually implement the technology.

Finally, what level of partnership is appropriate? If the start-up is developing multiple technologies, it is important to consider which ones to pitch to a strategic partner. Additional considerations occur if the potential partner has several different business units. While getting buy-in from multiple business units within a company can help accelerate a relationship, making multiple pitches can also hurt a start-up's chances. First, it can make the start-up seem less focused. Second, if one business unit likes one technology but another unit does not believe another of the start-up's technologies will work, the start-up may have inadvertently caused a potential conflict within the strategic partner. While rare, it is important you do your homework on the industrial partner to maximize your chance of success. This can be achieved by first finding a "champion" within a company and using them as a resource.

9.6.2 Industrial Perspective

The industrial company must undertake the task of developing a partnership strategy. This strategy should define the needs they are seeking to satisfy, the engagement models they are willing to employ, and which resources will be made available to partnerships. With this strategy in place, the industrial company can now initiate efforts to identify possible partnerships and to communicate their strategy through relevant networks in order to create opportunities. Creating this "deal flow" increases the probability of finding the right partnership.

After confirming the strategic alignment of a potential partnership, focus shifts to evaluating the feasibility of the partnership to address a specific need. The feasibility analysis can generally be performed without an NDA/CDA as it is focused on identifying the major risks associated with a technology or product. The feasibility analysis is usually performed by a team with considerable expertise representing a broad range of competencies. Any major risks identified at this stage do not necessarily stop a partnership but can be used to shape the scope of the efforts within the partnership. To illustrate major risk identification as part of a feasibility analysis, consider the following scenario:

> A recently founded start-up has spun out of university research efforts that demonstrate the ability to produce a novel sustainable solvent via the catalytic conversion of ligno-cellulosic material. The lab studies indicate the catalyst has a short lifetime and recovers activity well after regeneration.

During the feasibility analysis of the technology proposed in this scenario, it may be reasonable to conclude that a fluidized bed reactor is necessary for scale-up and that catalyst costs may contribute significantly to operating costs. These risks may now be used to define partnership targets. Example targets may include achieving a 10-fold increase in catalyst lifetime, reducing the catalyst manufacturing cost by 10%, or increasing yield of the solvent by 5%. This begins to define a scope for the partnership and metrics to measure progress.

After establishing strategic alignment and feasibility of the potential partnership, the viability of the partnership is analyzed next.

This stage intends to answer questions:

- Are both partners capable of performing the anticipated tasks within the partnership?
- Is the industrial company willing to make the needed resources available to the partnership?
- Is the engagement model sought by the start-up aligned with the industrial company's strategy?
- Is the partnership appropriate for the maturity of the start-up?
- Is there commitment of the industrial company's leadership to the principles of the partnership?
- Can agreeable terms for a partnership be obtained?

By answering these and other similar questions, the industrial company now has a complete view of the partnership scope and viability. The process to approve the partnership and finalize negotiations can proceed, leading to the kickoff and execution of the partnership.

9.7 Establishing the Partnership

For the industrial company establishing the partnership, it is vital to consider the culture of the start-up, pace of development, and scale at which the technology has been demonstrated. For the start-up, it is important to understand investor expectations for the partnership. Are they long-term investors that want to see a final product advance to market, or do they expect a relatively quick turnaround from their investment? Is there an expectation that if the partnership goes well an investment or buyout is possible? Setting clear expectations before the agreement is signed can go a long way in allaying frustrations down the road.

The aforementioned barriers also affect the process for establishing a strategic partnership. If the strategic partnership is approved, it can be structured such that the ideation, development, and validation phases occur within the partnership and can therefore be subject to different controls than those existing within the industrial company. This is not to suggest the pursuit of inappropriate behavior or the elimination of controls but rather that the needs of the larger company be addressed during the establishment phase in a manner that provides greater freedom to the partnership. By allowing the partnership to develop its own way of working, compliant with applicable regulations, the partnership can be a source of new ideas and business models that may have been suppressed within the larger industrial company.

During the establishment phase of a strategic partnership, both parties are looking to build an effective way of working together. Key to this is agreeing to milestones that focus on the near-term requirements. In addition, the teams should specify opportunities for scenario reevaluation and partnership adjustment that can help ensure a level of control that remains dynamic for both partners throughout the project.

When establishing a partnership, the importance of KPIs and other metrics to operational excellence is well established [18]. This requires the industrial company to address (i) what to measure and (ii) how to measure. Deciding what to measure depends on the types of partnerships that the industrial company engages in and what are the desired outcomes. Measuring the impact of partnerships via KPIs and metrics in the following areas should be considered [19]:

- Enterprise accountability (e.g. risk-adjusted return on capital, net promoter score)
- Customer relationships (e.g. brand awareness, new product adoption)
- Workplace analytics (e.g. collaboration through partnerships, effect of partnerships on employee engagement)
- Partner performance (e.g. project management metrics, idea generation)

9.8 Executing the Partnership

During the execution of a strategic partnership, good communication and risk management practices are critical [20] and benefit from the application of project management principles. The specific methodology employed (e.g. lean, agile, or waterfall) will depend on the nature of the project and preferences of the partners [21]. For example, an agile approach may be best suited to the validation of a new sustainable chemical product in applications where adaptability and iteration with customer feedback are most important. Alternatively, the more traditional waterfall methodology (where the project is carried out in distinct, well-defined stages) can help ensure the construction of a pilot plant meets the timeline and cost objectives.

In addition to managing the communication within the partnership, it is important to communicate the status to the leadership sponsor and update internal KPIs and metrics. Maintaining current metrics can help identify problems early in their development, potentially mitigating negative outcomes. Continued engagement of leadership will also help ensure resource availability and responsiveness.

To evaluate KPIs effectively, the challenge of obtaining useful data often arises. Existing data sources may suffice for some KPIs, but others could require integration into new data sources, potentially including the start-up. The additional integrations may not be an activity that the information technology areas have managed before and could have additional considerations regarding data security and privacy such as the recent European Union General Data Protection Regulation (EU GDPR) [22]. While these organizational disruptions are certainly manageable individually, when taken together they can become a significant challenge requiring resources in addition to those needed for the partnership.

Finally, the industrial partner should periodically reassess the partnership in a manner similar to that done in the evaluation phase. For partnerships with a time horizon of less than one year, significant changes to the scenario are less likely, while those that will continue for multiple years are nearly certain to encounter a change. Scenario changes might include the following:

- Change in resource availability for either partner
- Change in leadership commitment
- Identification of new risks that significantly alter the risk–benefit balance
- Strategy pivots for the industrial partner
- Target market, application, or technology pivots for the start-up

Similar to the establishment phase, these scenario changes may not necessarily force the end of the partnership. Reevaluation can help to ensure that the appropriate effort is put forth by the industrial partner.

9.9 Closing the Partnership

Most partnerships will eventually reach a conclusion, whether it is at the planned end point or due to a change in the scenario. Once development and validation are complete, the partner may also elect to license the technology or acquire the start-up outright. If these latter scenarios are not in play and the partnership is to end, the first point of attention is finalizing the project deliverables and holding a project debrief. The deliverables should have been established in the scope of the partnership, and now satisfaction of agreed upon terms should be verified. In the event that the terms of a deliverable are not satisfied and relief is unavailable under other terms of the agreement, legal council may be warranted. However, the hope is that this can be avoided through active management throughout the partnership.

The project debrief provides an opportunity to review the KPIs and metrics that were monitored throughout the partnership and to provide mutual feedback. Feedback should highlight what efforts were successful and the challenges that were encountered. The successful efforts may help to define best practices for both parties when engaging in future partnerships, while the discussion of challenges in a less stressful environment can reveal areas for continued improvement and focus.

The deliverables achieved by the partnership, the successful efforts, and the challenges should all be communicated to the leadership and more broadly within the industrial company. This communication is an opportunity to celebrate success, share knowledge, and make progress toward any cultural adaptations desired.

Finally, during the closing stage of the partnership, the possibility of continuing the partnership may arise. It can be tempting to simply agree to adjust the scope of the existing agreement to avoid aspects of negotiation and approval, but this is not recommended in most cases. The closure and then development of a new scope provides both parties with the opportunity to systematically and thoroughly evaluate what efforts should be prioritized. This time is also an opportunity to consider the original agreement terms with the results from the partnership and adapt if necessary. Additionally, the learnings from the project debrief can be explicitly integrated into the new scope thus avoiding a loss in knowledge capture. A new agreement may come at a loss of momentum but will lay the groundwork for continued success of the partnership.

9.10   Case Studies

Case Study 9.1

Siluria Technologies and Braskem

The players:

- Siluria Technologies, a start-up headquartered in San Francisco, California, developing the oxidative coupling of methane to produce olefins
- Braskem, a global petrochemical company headquartered in Sao Paulo, Brazil, with operations in Brazil, North America, and Europe

Motivation: Siluria's technology reached the development stage where scale-up to a demonstration plant with realistic feeds and products was needed. Braskem, being a large consumer of olefins, had interest in diversified feedstock supplies.

The partnership: The two companies negotiated a partnership in 2014 that allowed Siluria to gain access to Braskem's La Porte manufacturing site to construct a demonstration plant. Additionally, the agreement included the validation of the olefins produced by Siluria's technology with Braskem's operations on-site and set forth a framework for commercial agreements in the future [23].

The outcome: Siluria was able to access existing Braskem infrastructure to demonstrate its technology. A demonstration plant was successfully constructed and began operation in 2015 [24]. A series of campaigns were run on the demonstration plant [25], which attracted the attention of McDermott International, a global engineering, procurement, and construction company headquartered in Houston, Texas. In 2019 Siluria entered into an asset purchase agreement with McDermott to facilitate the commercialization and distribution of the oxidative coupling technology [26]. Braskem gained direct experience using the technology and the rights to potentially access the technology in the future with favorable terms.

Case Study 9.2

Novomer and Albemarle

The players:

- Novomer, a start-up headquartered in Boston, Massachusetts, developing a carbonylation catalyst technology to produce polyols from propylene oxide
- Albemarle, a global specialty chemical company headquartered in Baton Rouge, Louisiana, with a focus on pharmaceutical, agrochemical, and specialty materials

(Continued)

Case Study 9.2 (Continued)

Motivation: Novomer had successfully completed a US Department of Energy Phase One project to perform experiments and generate a preliminary pilot plant design for the production of polypropylene carbonate. It was selected for a subsequent Phase Two award to fund scale-up of the technology and was thus looking for an appropriate scale-up partner. Albemarle operated several flexible manufacturing plants and had an interest in developing new sustainable specialty chemicals.

The partnership: In 2010 the companies entered into a partnership to execute the Phase Two scale-up by modifying Albemarle assets to incorporate the Novomer technology [27].

The outcome: In 2013 the first large-scale manufacturing run of polypropylene carbonate was completed, producing 7 tons of product to accelerate validation in the polyurethane value-chain [28]. This attracted the attention of Saudi Aramco, a national integrated energy company headquartered in Riyadh, Saudi Arabia. A new brand was ultimately created, called Converge, which was spun out of Novomer and acquired by Saudi Aramco in 2016. It was valued up to $100 million [29]. The polyol brand Converge is actively marketed by Saudi Aramco with plans to construct a full-scale facility in Saudi Arabia [30].

Case Study 9.3

Itaconix and Nouryon (AkzoNobel Specialty Chemicals)

The players:

- Itaconix PLC, a publicly traded company founded in 2008 in New Hampshire, developing products from the novel ability to efficiently polymerize itaconic acid
- Nouryon, a global specialty chemical company that separated from Albemarle in 2018, headquartered in Amsterdam, The Netherlands

Motivation: Itaconix successfully demonstrated the ability to polymerize itaconic acid, derived from sugars via fermentation, and apply these polymers in a broad range of applications. AkzoNobel had established a sustainability initiative that included the use of renewable ingredients and was seeking solutions for the markets it serves.

The partnership: In 2017, the companies entered into a joint development to explore the development and commercialization of the polymers produced using the Itaconix technology [31].

The outcome: In 2019, Nouryon signed a supply agreement with Itaconix to supply products into the detergent and cleaner markets [32]. Shortly thereafter Nouryon signed a second exclusive supply agreement with Itaconix for the personal-care market [33]. The joint development provided critical market access to validate the performance of the Itaconix materials and established a supply-chain integration to expand the reach and scale of operation.

References

1 Deloitte Consulting GmbH (2017). *Beyond the Noise: The Megatrends of Tomorrow's World*. LOGOPUBLIX Fachbuch Verlag.

2 Meister, J.C. and Mulcahy, K.J. (2017). *The Future Workplace Experience: 10 Rules for Mastering Disruption in Recruiting and Engaging Employees*. New York: McGraw-Hill Education.

3 International Council of Chemical Associations. (2017). Global chemical industry contributions to the sustainable development goals. https://www.icca-chem.org/wp-content/uploads/2017/02/Global-Chemical-Industry-Contributions-to-the-UN-Sustainable-Development-Goals.pdf (accessed 12 November 2019)

4 ExxonMobil. (2017). 2017 sustainability report highlights. https://corporate.exxonmobil.com/en/~/media/Global/Files/sustainability-report/publication/2017-Sustainability-Report.pdf (accessed 31 October 2020).

5 BASF. (2018.) BASF report 2018: economic, environmental and social performance. https://report.basf.com/2018/en/servicepages/downloads/files/BASF_Report_2018.pdf (accessed 9 November 2020).

6 DuPont de Nemours. (2019). Sustainability roadmap. https://www.dupont.com/content/dam/dupont/amer/us/en/corporate/about-us/Sustainability/DuPont_Sustainability_Roadmap_20190107.pdf (accessed 31 October 2020).

7 Kirsner, S. (2018). The biggest obstacles to innovation in large companies. *Harvard Business Review* (30 July 30). https://hbr.org/2018/07/the-biggest-obstacles-to-innovation-in-large-companies (accessed 31 October 2020).

8 KBC and International Quality & Productivity Centre. (2017). Operational excellence in refining and petrochemicals. https://www.kbc.global/insights/whitepapers/operational-excellence-in-refining-and-petrochemicals/ (accessed 9 November 2020).

9 Arena, M., Cross, R., Sims, J., and Uhl-Bien, M. (2017). How to catalyze innovation in your organization." *MIT Sloan Management Review*. (13 June 13). https://sloanreview.mit.edu/article/how-to-catalyze-innovation-in-your-organization (accessed 31 October 2020).

10 Citrine. (2018). BASF and Citrine Informatics collaborate to use artificial intelligence to develop new catalyst technology. Press release (21 June). https://citrine.io/media-post/basf-and-citrine-informatics-collaborate-to-use-artificial-intelligence-to-develop-new-catalyst-technology (accessed 31 October 2020).

11 BASF. (n.d.). Digitalization. https://www.basf.com/global/en/who-we-are/digitalization.html (accessed 27 October 2019).

12 Page, T., Rahnema, A., Murphy, T., and McDowell, T. (2016). Unlocking the flexible organization: organizational design for an uncertain future. Deloitte Development LLC. https://www2.deloitte.com/content/dam/Deloitte/global/Documents/HumanCapital/gx-hc-unlocking-flexible-%20organization.pdf (accessed 31 October 2020).

13 DIC Corporation. (2018). DIC Corporation Signs JDA with Checkerspot to Develop Advanced High-Performance Materials. Press release (8 May 8). https://www.dic-global.com/en/news/2018/ir/20180508000010.html (accessed 9 November 2020).

14 Catino, E. (n.d.). The green ski revolution: Matt Sterbenz launches the WNDR Alpine ski brand." *Freeskier*. https://freeskier.com/stories/the-green-ski-revolution-matt-sterbenz-launches-the-wnder-alpine-ski-brand (accessed 31 October 2020).

15 Tullo, A. (2018). Dow Chemical wins $1 billion judgement against Nova Chemicals." *Chemical & Engineering News* (28 June)/. https://cen.acs.org/business/petrochemicals/Dow-Chemical-wins-1-billion/96/i27 (accessed 31 October 2020).

16 Mehlman, S.K., Uribe-Saucedo, S., Taylor, R.P. et al. (2010). Better practices for managing intellectual assets in collaborations. *Research-Technology Management* 53 (1): 55–66. https://doi.org/10.1080/08956308.2010.11657612.

17 Chesbrough, H.W. and Brunswicker, S. (2013). *Managing Open Innovation in Large Firms: Survey Report; Executive Survey on Open Innovation 2013*. Stuttgart: Fraunhofer-Verl.

18 Mitchell, J.S. (2015). *Operational Excellence: Journey to Creating Sustainable Value*. Hoboken, NJ: Wiley.

19 Schrage, M. (2018). Five categories to focus your KPIs." *MIT Sloan Management Review Strategic Measurement* (21 September 21). https://sloanreview.mit.edu/article/five-categories-to-focus-your-kpis (accessed 31 October 2020).

20 Slowinski, G. and Sagal, M.W. (2010). Good practices in open innovation. *Research-Technology Management* 53 (5): 38–45. https://doi.org/10.1080/08956308.2010.11657649.

21 Pitagorsky, G. (2006). *Agile and Lean Project Management: A Zen-like Approach to Find Just the 'Right' Degree of Formality for Your Project*. Seattle, WA: Project Management Institute.

22 EUR-Lex. (2018). Communication from the Commission to the European Parliament and the Council. (24 January). https://eur-lex.europa.eu/legal-content/EN/TXT/?qid=1517578296944&uri=CELEX%3A52018DC0043 (accessed 31 October 2020).

23 Siluria. (2014). Siluria Technologies announces construction and site selection for its OCM demonstration unit in La Porte, Texas. Press release (15 January). https://www.globenewswire.com/news-release/2014/01/15/1130192/0/en/Siluria-Technologies-Announces-Construction-and-Site-Selection-for-Its-OCM-Demonstration-Unit-in-La-Porte-Texas.html (accessed 9 November 2020).

24 Siluria. (2015). Siluria Technologies announces successful start-up of world's first demonstration plant to directly convert natural gas to ethylene. Press release (1 April 1). https://www.prnewswire.com/news-releases/siluria-technologies-announces-successful-start-up-of-worlds-first-demonstration-plant-to-directly-convert-natural-gas-to-ethylene-300059536.html (accessed 9 November 2020).

25 Siluria. (2016). Siluria announces one year of successful operations of its disruptive ethylene technology. Press release (11 May). https://www.prnewswire.com/news-releases/siluria-announces-one-year-of-successful-operations-of-its-disruptive-ethylene-technology-300266816.html (accessed 9 November 2020).

26 Tullo, A. (2019). McDermott buys Siluria for oxidative methane-coupling technology. *Chemical & Engineering News* (9 August).

27 Wilkinson, M. (2010). "ecycling CO_2 to make plastic. Chemistry World (28 July)". https://www.chemistryworld.com/news/recycling-co2-to-make-plastic/3003416.article (accessed 9 November 2020).

28 US Department of Energy, Office of Fossil Energy. (2013). Recycling carbon dioxide to make plastics (20 May). https://www.energy.gov/fe/articles/recycling-carbon-dioxi (accessed 9 November 2020)

29 Tullo, A. (2016). Aramco buys Novomer's CO_2-based polyols business. Chemical & Engineering News (4 November). https://cen.acs.org/articles/94/web/2016/11/Aramco-buys-Novomers-CO2-based.html (accessed 9 November 2020).

30 Saudi Aramco. (2016). Saudi Aramco acquires Novomer's polyol business and associated technologies, enhancing its downstream expansion strategy. Press release (3 November 3). https://www.saudiaramco.com/en/news-media/news/2016/acquires-novomers-polyol-business-downstream-expansion (accessed 31 October 2020).

31 Nouryon. (20197). AkzoNobel signs cooperation agreement on bio-based polymer technology with Itaconix. https://www.nouryon.com/news-and-events/news-overview/2017/akzonobel-signs-cooperation-agreement-on-bio-based-polymer-technology-with-itaconix (accessed 31 October 2020).

32 Nouryon. (2019). Nouryon signs deal with Itaconix for bio-based polymers for the detergents market. https://www.nouryon.com/news-and-events/news-overview/2019/nouryon-signs-deal-with-itaconix-for-bio-based-polymers-for-the-detergents-market (accessed 31 October 2020).

33 Nouryon. (2019). Nouryon to supply new bio-based polymers to the personal care market. https://www.nouryon.com/news-and-events/news-overview/2019/nouryon-to-supply-new-bio-based-polymers-to-the-personal-care-market (accessed 31 October 2020).

10

Bridging the Gap 1: From Eureka Moment to Validation

Peiman Hosseini[1] and Harish Bhaskaran[2]

[1] *Bodle Technologies, Oxford, UK*
[2] *Dept of Materials, University of Oxford, Oxford, UK*

10.1 Introduction

This chapter will present concepts and examples regarding planning, managing, and executing a technology development plan with an early-stage company. The specific techniques presented here are by no means the only ones available, but rather a small collection of processes that proved effective in the early years of our establishing and growing Bodle Technologies Ltd.

Due to the nature of the products we are commercializing, this chapter admittedly takes a more engineering-centric approach. However, all the concepts presented are directly applicable to those encountered when developing any sustainable chemical technology. You are encouraged to find, modify, and use whatever approach is most appropriate for your company.

As background, Bodle is pursuing the commercialization of a new type of reflective display technology based on a special class of chalcogenide glasses known as phase change materials (PCMs) [1, 2]. PCMs are functional materials characterized by two distinct phases: crystalline and amorphous. Both phases are stable at room temperature yet have very different optical and electrical properties. Interestingly, one can switch between the two phases an almost infinite amount of times using very short energy pulses. Optical storage technologies like DVD-RW and Blu-ray use lasers to toggle between the two phases and store data as localized bits on a PCM thin film [3].

We discovered the science behind Bodle in 2013 while trying to understand the fundamental relationship between the electrical and optical properties of these unique materials. I remember running a simulation using a technique learned a few hours before from one of our colleagues at Oxford, when an unexpected color-changing effect suddenly "appeared" on my screen. I spent the following hours searching through the literature trying to find a publication that described this unusual opto-electronic effect, without success.

We later understood how that simple color change was in fact an entirely new way to generate and modulate flat surfaces using simple electrical pulses. When we were simulating these optical-changing properties, none of us knew much about optical coatings (you could

How to Commercialize Chemical Technologies for a Sustainable Future, First Edition.
Edited by Timothy J. Clark and Andrew S. Pasternak.
© 2021 John Wiley & Sons Ltd. Published 2021 by John Wiley & Sons Ltd.

say we knew absolutely nothing), which is probably why we ran simulations on extremely thin films (5–10 nm) on which no expert would have wasted their time. Visible light has a wavelength of roughly 400–700 nm, our films were less than a tenth of the shortest wavelength of light. Intuitively, one might expect to see near-zero optical modulation using an active material so thin. What we found was the exact opposite: using thin layers not only resulted in outstanding light modulation properties, but the expected contrast was considerably higher for the thinnest of the films.

This was counterintuitive to experts, but we were not experts, and we ended up discovering something rather remarkable. We published our results in *Nature* in 2014 with many news media outlets talking about it as the next big thing in displays [4, 5]. Today, Bodle's core expertise and technology offering lies on a special type of extremely thin optical coating, one that can be modulated and tuned at will using both electrical and optical pulses. Applications span from full color, reflective displays to security films to novel energy-efficient smart glazing. More importantly, a new device that can manipulate light without the need for constant power offers sustainability benefits in all of the aforementioned applications. For example, in smart glazing applications, bistable coatings tailored to the infrared spectrum can control the amount of heat transmitted through a window depending on the season. This greatly reduces the energy required for temperature management.

10.2 Fundamental Research Leading to an Invention

It all starts with an idea and a few results. As scientists, we have the great privilege of devoting ourselves to the noble quest of developing the scientific understanding of the universe. Many of today's greatest innovations came from regular scientists trying to study and replicate interesting phenomena in their lab. The vast majority of the time, these phenomena might have great scientific importance but no immediate commercial applicability. However, from time to time, a PhD student or postdoc discovers something that indeed has some clear commercial relevance with an appropriate market identified.

The question of how one can foster more commercially relevant research remains mostly unanswered, although recent work has shown that it is often freedom to pursue your creativity in areas beyond the scope of a specific project [6]. How do you identify a market for your scientific discovery? As described in Chapter 2, one of the most effective ways is to talk to as many potential customers as possible, a process often called "customer discovery."

Finding the right customer or strategic partner before looking for funding will give you a tremendous advantage. By leveraging this potential customer or partner's expertise, a development roadmap and the resources needed for its execution can be planned in a more professional and lean way. Most universities now have some level of support for the aspiring entrepreneur to build a business case and protect the intellectual property (IP) for their discovery as well as grants and office space. Chapter 5 fully details the support that exists in various regions' entrepreneurial ecosystems.

When we started Bodle, explaining to investors the advantages, and differentiating characteristics of a new display technology was relatively simple. In the era of smartphones and flat-screen TVs, everyone knows what a display is and understands the need for a better one. Most people have grown accustomed to that sector's innovation cycle with larger, better, cheaper products available every year. Yet, in spite of this, it took 18 months to raise our first round of investment!

The challenges that we faced were different and unique. Here are a few examples of important and informative questions asked by our future investors in 2015. Our responses to these questions shaped the business and eventually allowed us to raise that first round.

- *What is your business plan and your path to initial revenues?*
  Our business plan changed over the years, but it was initially a licensing model. We would license our patents and know-how portfolio to large display manufacturers in Asia in exchange for fees and royalties. Our answer was immediately followed by a second question.
- *Licensing businesses are often not very exciting. Could you manufacture and sell displays yourself?*
  Display factories are getting bigger and bigger. $4 to 5 billion of capital is required to build a new display factory, and China is leading the world. A small start-up from the UK is unlikely to have access to billions of pounds, so a licensing model seemed the only sensible solution. To make the business more appealing, we considered the introduction of the same technology to different markets, versioning it, making it flexible, and so on. These are common and effective tactics to increase the value of a start-up with an IP-rich technology, but with large capital requirements for direct manufacturing.
- *Do you have any patents? Who owns the IP?*
  Before we published the fundamental paper describing our technology, we filed several provisional patents on all of the possible applications we could think of. This was important and allowed us to retain the value of the research undertaken at the university. When we spun out Bodle, the initial IP was licensed exclusively to us with full options to sublicense to third parties. Despite potentially being viewed as a hindrance, this scheme often protects an inexperienced founder from future hostile takeovers by refusing a full assignment to a party that has no real intention of commercializing the technology. Typical examples might be patent trolls or unfriendly competitors who simply want to terminate the technology. The start-up company is always given the option to obtain a full assignment in case of a genuine exit event such as an acquisition or initial public offering (IPO).
- *How big is your addressable market?*
  The display market size is on the order of $120 to $150 billion. The vast majority is dominated by LCD and OLED displays that we were not looking to displace. Those technologies are extremely mature, relatively inexpensive, and almost impossible to compete against. We aimed to create new markets that did not exist at the time. Bodle had a goal of making any surface that does not shine light a screen. It uses ambient light to convey information – a moving, colorful magazine. No other technology can do that today.
- *What is your exit strategy? How long for a return?*
  Our exit strategy at the time was an acquisition by a large company. Developing a new display technology takes many years; the most successful ones took at least 10 years and hundreds of millions of dollars to reach the market. From the beginning, it was clear that we needed the support of an understanding investor with a long-term view.

After roughly 18 months of pitching and negotiations, our first Seed investment round was transferred to our bank account. Shortly thereafter, we started hiring engineers and buying much needed equipment to further develop the technology.

10.3 Proving the Concept

Identify main technical challenges early. Apply agile approaches to "firsts." If no one has done this before, can you demonstrate what is feasible with a quick-and-dirty experiment? De-risk the proposition.

Transformational technologies do not simply appear on the market. Large teams of tens, hundreds, or even thousands of individuals work tirelessly designing every functionality, component, and interface to deliver a product that adds true value to the market. One might ask, how did they know they would succeed? Why would someone invest millions of dollars in a venture that no one has ever pursued before?

The answer is that those individuals had to prove that their technology had legs well before they were able to raise millions of dollars from venture capital and corporate investors. This initial phase is normally called the proof-of-concept or "Seed" stage of a new venture.

You have to think laterally when validating a new technology. How can you prove that an idea will work without adequate resources, people, or lab space? Most importantly, how can you do it in a short amount of time? All you have at the beginning are a couple of journal papers and a patent filing or two, just like we had back then. It might not seem like much, but it is objectively a perfect start. You made a solid discovery, identified an initial market, landscaped your potential competitors (see Chapter 2), and have a vague idea of your first product.

The logical next step is writing down all the technical challenges you can think of. Ideally, you should do this exercise as a brainstorming session with your co-founders, colleagues, and trusted friends. As is the case with every technical founder, you will probably have already thought of the challenges and potential solutions to them. Now is the time to write them down and share them with your colleagues. Start with a white board, write everything you can think of, and try to tear your project down to the nuts and bolts. Think about what you are trying to demonstrate. What are the unique characteristics of your technology, and how can you prove them quickly with some simple experiments?

Any newly developed material or technology often requires integration with some existing system to show its full capabilities. When we started Bodle, it was clear that our pixels were only part of a display system made of a backplane, chip-on-glass driver, driving algorithms, and so on. Each one of these components needed to be understood, adapted, and integrated to measure the true performance of our display technology. Unfortunately, we simply did not have the resources nor the time to create a full working display, and that was fine as we did not need to.

No sophisticated investor expects you to have a ready-to-ship product at the Seed stage. Your main concern should be to *de-risk the technology as much as possible while identifying all the foreseeable challenges that lie ahead.*

The next key question is: What is the absolute minimum that has to work for the whole proposition to make sense? Can you demonstrate it with some quick and dirty experiments in the lab? Remember the objective is to demonstrate the validity of your business proposition, not to publish a paper in a scientific journal. In the early days of a company, the team is composed of essentially the co-founders and perhaps one or two believer-scientists that the founders have convinced to join the project. Scarcity of time and resources will likely put a lot of strain on everyone; it is imperative that the team proves the proposition in the shortest time and most effective way possible. The key at this stage is not to make perfection the enemy of the possible – especially true for academic founders who tend to value perfection. As much as investors and everyone else will have you believe, there is no such thing as a perfect demonstration that immediately leads to a blockbuster product.

10.4 The Tech Team: Moving Beyond an Academic Group

Like a small research group but different.

When we started Bodle, we quickly realized that a solitary, postdoc-type mentality would not result in a successful high-tech company. As a student or postdoc, most of your work is carried out independently in your principal investigator's lab. You are often encouraged to explore your area of study freely, looking for interesting phenomena that might nudge the boundaries of your research a little further. Critically, it often does not matter what that knowledge really is as long as it is novel, interesting, and achievable. This mentality quite rightly pushes a scientist to explore a variety of ideas at the same time, understand their feasibility, run experiments, collect results, write a paper, and move on. Ideas that did not quite make it because they were too difficult, plain wrong, or simply not interesting enough are dropped or put on the back burner for future exploration.

Although an early-stage start-up has many similarities to an academic research group, there are some fundamental differences that we believe are critical to highlight. A company has first and foremost a *commercial focus*; that is the sole purpose of its existence. There is nothing wrong when pure scientific knowledge emerges as a by-product of a commercial enterprise, but that is not the primary objective. The output is also radically different; open literature papers and presentations make space for patents, know-how, demonstrators, and scalable products.

The mentality of a start-up scientist is also radically different from that of an academic scientist. The business might require a technological breakthrough to justify its existence, and that should be the sole objective of the technical team. Pivoting to a different project might still be possible at the early stages of the company but will become exponentially more difficult as time passes. The takeaway message is that no one wants to invest in a company that does not know what it was created for. Uncertainty and a lack of focus often lead to wasted resources (time, money, reputation, etc.), and most investors will see this behavior as an indicator of much bigger problems and greater risk. The company should always unambiguously state its mission (what it was created to do) and its vision (what it intends to become in the future). The CEO and senior management of the company should always refer to those statements when presented with a choice that will greatly impact the future of the company.

It is important to remember that the notion of proof-of-concept extends far beyond that of a simple technology demonstration. Sophisticated investors will also scrutinize the founding team's behavior and execution during this phase to ensure that they deliver in the leanest way possible. The founders will also set the culture of the company for the foreseeable future. Are they competent? Do they possess the ability to operate under stress? What is their vision for the company? It might be a stressful time, but it should also be an exciting one. An early-stage start-up is first and foremost the projection of the vision of its founders. Strip the company of all its assets (funding, resources, processes, products, etc.), and what is left are the founders, their vision, and the culture. This is true even when a professional executive team is parachuted in by investors at great cost. They can help professionalize the customer relationships but cannot substitute for founder vision, enthusiasm, and belief in the technology.

10.5 Developing the Road Map

Your company has finally passed the Seed stage and now has enough money to hire more people and buy a few much-needed gadgets for the lab. Congratulations – you are now in a fairly exclusive club! Now what? Now it must all change, and you must change, again and quickly! You (and your company) have already evolved once, from a few solitary scientists to a tiny team of misfits with a can-do attitude. The next step is to evolve once again into something that more closely resembles a real company. Your immediate goals are likely to be adding people and giving them milestones, responsibilities, and deadlines. Call it creating a structure.

Where to start? You need a map – a mental map of all the things that you will need to develop in the coming months. As with every plan, it can and will change almost every day, and that is OK.

Several mind mapping techniques have been described over the years. We chose a particular type of "causal" diagram known as Ishikawa or Fishbone diagrams. Fishbone diagrams were popularized by Kaoru Ishikawa in the 1960s while working with the Kawasaki shipyards quality control department. Historically, the fundamental scope of a Fishbone diagram was the identification of cause-effect relationships with the main "effect" at the top resulting from the "causes" listed on the diagram. While usually employed for root-cause analysis and quality control, this mapping technique can be repurposed as an effective tool for high-level definition and implementation of research and development projects.

Whatever method you use, visualizing a project as a whole, and from a very high-level perspective, has several unexpected benefits that will help the technical team identify gaps and duplicates in the development plan. This is especially helpful at the beginning of a start-up's life cycle where the team does not have experience working together on a particular project and are likely to underestimate the sheer volume of all the components and activities needed to achieve the company's objectives.

Figure 10.1 shows an example of a fishbone diagram from early 2016, the year Bodle recruited its first technical team in the UK.

From a general point of view, an "effect" can be anything the company is looking to achieve. In our case we were trying to demonstrate the basic building block of our technology, one single pixel switching on and off several times. The "causes" leading to the successful demo were divided into these seven larger groups:

- People: The engineers and managers involved in the project
- Materials: The new materials developed specifically for the project
- Fabrication: The microfabrication processes necessary to generate the pixels

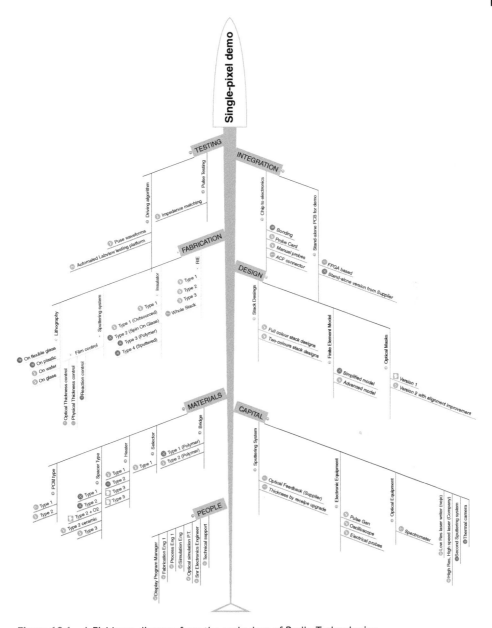

Figure 10.1 A Fishbone diagram from the early days of Bodle Technologies.

- Testing: Protocol and equipment required to evaluate and score the results
- Capital: Physical equipment needed to fabricate and evaluate the devices
- Design: Software, CAD files, and everything else needed in the design phase
- Integration: System-level integration between single pixels and external electronics

The diagram has additional icons (checkmark, thumbs down etc.) next to each activity that outline its status with respect to the development plan. As discussed previously, the aim of a Fishbone diagram is not to schedule, prioritize, or assign tasks but rather to visualize the entire project on a single piece of paper. Once your diagram is complete, try to answer a few questions together with your team:

1. Do you have everyone and everything you need to demonstrate and de-risk your technology proposition?
2. Have you focused the right amount of attention and resources to the core part of your technology? How much energy is "wasted" on peripherals that are useful (nice to have) but not critically important (need to have) to your proposition?
3. Could you partner with research institutions, service providers, or strategic partners (see Chapter 9) to share some of the heavy lifting? Sharing efforts may require sharing some of the future rewards too, unless it is transactional. Is this something your company could live with? It is always a good idea to keep the most valuable part of the core technology development internal to avoid damaging the overall business proposition.

10.6   Defining Your Technology Development Requirements

Developing a new, disruptive technology from scratch without any historical design precedent to use as a blueprint is undoubtedly one of the hardest challenges when commercializing a sustainable chemistry technology. The early phase of defining the technology requirements before moving to the design and simulation phase is arguably the most consequential step of the entire development process.

When an R&D team starts defining the requirements of a new project, it might seem appropriate to treat the technology as a stand-alone module separate from the rest of the system. This fallacy in treating the problem as an isolated component is almost always a recipe for disaster. New technologies are unlikely to be "plug and play"; existing and adjacent components will not be ready to accommodate them as they were designed by a different team with different objectives at some point in the past. When separate teams develop single components in isolation and without a system overview of the entire project, issues will very likely arise during the integration phase. A more appropriate approach is to acknowledge the complexity of the design phase from the outset and pay special attention to the interfaces between your technology and the rest of the system. This system engineering–inspired approach to R&D has the capability to visualize and highlight hidden interdependencies and feedback loops that often emerge from introducing a new technology into an existing solution.

Complex systems [7] can be especially hard to manage in terms of resource allocation and design workflow. When many components have interdependencies and connect to each other in several ways, where should the design process even begin? It is easy to understand the inadequacy of a simple serial approach where interdependencies and feedback loops give each component equal importance, making it difficult to decide which component should have priority and which ones could be designed in parallel.

Fortunately, there are a few tools specifically for managing complex systems that help simplify the job of the R&D team. Design structure matrix (DSM) is a state-of-the-art management technique that "unwraps" an intricate project into small chunks of work

Table 10.1 An example of design structure matrix (DSM) before sorting. Input activities are columns, output activities are rows.

	#	A	B	C	D	E	F	G	H	I	J	K	L	M
Demo Layout Design	A	■	x											
Target Definition	B		■	x		x								
Customer requirements	C			■										
Demo Simulation	D	x			■									
Market Analysis	E		x	x		■								
Demo Evaluation	F		x		x		■							x
Initiate Manufacturing	G							■	x		x			
Knowledge transfer	H	x	x				x		■		x			
Process definition	I	x			x					■				
Identify scale up partner	J		x	x							■			
Design for Manufacturing	K							x		x		■		
Process calibration	L									x			■	
Demo Fabrication	M	x								x			x	■

that can be executed as a sequence of activities. A brief example of DSM applied to a new semiconductor device concept is presented in Table 10.1 to help you understand the potential of this technique.

In this example, a start-up is planning to commercialize a new device by demonstrating a proof-of-concept device in-house before transferring to a scale-up partner for manufacturing. Table 10.1 shows the design matrix of all the steps that the company must complete in order to commercialize the technology. Each activity must be replicated on both axes (here shown as letters A to M) and has a number of inputs and outputs. Inputs are defined as everything that specific activity needs before it can be executed, while outputs are everything that is generated after the activity is completed. Examples of inputs are results from a previous analysis, experiment, or study. Examples of outputs are operating instructions for a new process or a series of specifications for a future product. Inputs are visualized as "x" in the columns, while rows are the outputs.

As an example, the "Knowledge transfer" output step (moving the know-how from the start-up to the scale-up provider) requires input from the "Demo layout design" and "Target definition" steps, the "Demo evaluation" results, and completion of the "Identify scale-up partner" step. Filling in a DSM table like the one presented here helps identify and map the interdependencies and iteration loops inherently present in every complex project.

From the input/output relationship matrix of Table 10.1, you obtain the results in Table 10.2 by applying a simple sorting algorithm. This is how you do it:

1. First, move any output activities with no required inputs (e.g. "Customer requirements," which does not have an "x" in its row) to the top of the table.
2. Output activities with direct inputs from those top ones follow systematically. Here "Market analysis" and "Target definition" are the first two to follow since they both need only the input from "Customer requirements" to be executed. "Demo layout design" goes next since it requires inputs from the "Market analysis" and "Target definition." This sorting technique continues until the new table is completed.

Table 10.2 An example of design structure matrix (DSM) after sorting. Input activities are columns, output activities are rows.

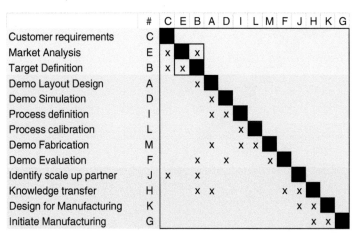

3. Once completed, the table should not show any "x" or relationship above the array's diagonal, as shown in Table 10.2.

The following are a few important details emerging from this example:

- The entire project is roughly divided into three steps, we will call them: definition (shaded blue), proof-of-concept (shaded green), and scale-up (shaded pink).
- "Customer requirements" does not require any input from the previous tasks, making it the first task that begins the project.
- "Market analysis" and "Target definition" are coupled tasks. The term "coupled" refers to two or more tasks having both inputs and outputs connecting *at the same time*. Coupled tasks are iterative and therefore executed in parallel with tight interaction between each other.
- "Identify scale-up partner" can be executed in parallel to the proof-of-concept tasks.

The tasks are now presented as a logical sequence of activities that can be scheduled using more traditional project management tools such as Gantt charts, critical path, critical chains, etc. Figure 10.2 shows an example of a Gantt chart, which is a type of bar chart used to illustrate a project's schedule, applied to this DSM example.

A full understanding of DSM is not in scope of this concise guidebook; however, more information can be found in the literature [8].

10.7 The Innovation Cycle: Design, Simulate, Fabricate, Test, Iterate

Previously, we introduced the importance of visualizing a project as a whole before assembling a team and setting up a plan to execute on it. Fundamentally, most projects are iterative

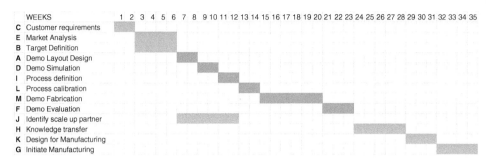

Figure 10.2 A corresponding Gantt chart of the DSM example.

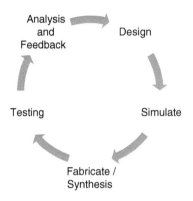

Figure 10.3 A typical development cycle for new technologies.

in nature. This is especially true for projects involving novel and disruptive technologies where a high degree of uncertainty makes it difficult to predict the behavior of a complete system a priori. Figure 10.3 depicts a typical development approach.

The technology team begins designing and simulating the new technology to the best of their knowledge based on the state of the art. The system is synthesized, and its performance measured and analyzed using a common framework defined at the beginning of the project.

The measurements are fed back to the design team who will create a new iteration based on the preceding evaluation with, ideally, improved performance, and the cycle continues until the target is ideally reached. Each iteration will provide learnings regarding the performance of the system, although not necessarily improvements. Adjustments may actually be detrimental, which in itself is valuable information. Indeed, when executing a newly defined development cycle, it is important to remember that "bad" results, defined as results that move further away from the specified targets, are just as important as "good" results. Identifying output trends from input design and process variables is the preferred outcome of any project. Future iterations should build on the knowledge gained throughout the entire project, not just the last iteration.

It is often tempting as methodical scientists to iterate by changing variables one at a time and measuring the resulting behavior sequentially. This is a slow and ineffective approach to product or process development. Design-of-experiment (DoE) techniques should be considered instead [9]. Briefly, a DoE approach uses applied statistics to conduct, analyze, and interpret controlled tests to evaluate factors (inputs) that control certain parameters (outputs). An input can be any variable that has significance to the system such as reaction temperature, reagent choice/loading, or process time. The outputs are measurable quantities that define the performance of the system (such as product yield, purity, mechanical integrity, etc.) and are typically being optimized.

10.8 Accelerating the Process

You have reached a point where your plan is executing at full speed, your team is familiar with the expected quality of the work, and the level of details needed in your documentation is acceptable (e.g. standard operating procedures, process definition and management system, etc.). Basically, everything seems to be in place, and you can concentrate on other aspects of the business while you wait for the outcome of the plan to come to fruition. As is often the case with many start-ups, time is always of the essence even when funding is not an issue. Can you accelerate the development in any way? Where are the bottlenecks, and how do you remove them? Visualizing workflow is an effective technique for identifying bottlenecks and modeling the effect of managerial modifications to address them.

10.8.1 An Example in Workflow Management

Imagine having five compositions of a family of new materials that you want to test. Each composition takes one day to design, one day to fabricate, three days to test, and two days to analyze the data and feed back to the design team. The simplest flow implementation is a *classical serial* first-in, first-out approach (Figure 10.4).

This simple workflow has some immediate pros and obvious cons.

Pros

- In principle, one well-trained chemist or engineer can run all aspects of the entire set of experiments.
- Efficient communication between activities is not required; the same person is doing everything!

Cons

- Resource and equipment utilization are very low as those needed for a particular activity stay idle while waiting for the previous to finish.

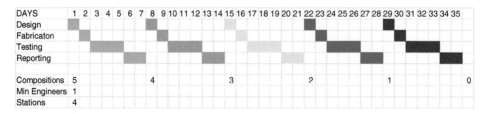

Figure 10.4 Classical serial approach to workflow management.

- The total time required to complete the task is high and increases proportionally with increasing sample numbers.

A slightly improved approach is a *staggered-serial approach* where each activity is constantly taking place without waiting for the previous one to finish (Figure 10.5). Using this technique increases your efficiency and can considerably reduce the total time required to complete the work.

Pros

- Resource and equipment utilization are close to 100%.
- The total time is now roughly dependent on the longest activity multiplied by the number of samples. This is markedly reduced in comparison to the classical serial approach.

Cons

- An increase in R&D staffing is required to complete the work, up to four times than needed with the classical serial approach.
- Related to the previous point, you need to seek out people with diverse backgrounds or demonstrated versatility when recruiting for R&D roles to ensure they are able to take on various workflows simultaneously.

Most R&D projects are usually a variation on one of these basic methodologies. Once the workflow has been visualized, it is important to identify the bottlenecks and constraints with the greatest impact on timing and scheduling.

Looking at the previous examples in Figures 10.3 and 10.5, it is clearly evident that testing is affecting the total time considerably. Ideally, you would like to accelerate testing and reduce the time it takes from three days to one day to eliminate idle time between activities. Here are a few potential ways to implement this:

- Automate the task. Testing the characteristics of a new material is often a repetitive and well-structured effort that has to be carried out with maximum diligence and precision. Automated testing is often a sensible solution that increases throughput and improves accuracy, efficiency, and traceability. However, it often requires a considerable amount of initial capital and should be carefully analyzed in terms of return on investment.
- Parallelize the task. Do you need any special equipment to test your samples? How expensive is it? Given your average daily financial burn-rate (the cash needed to run the entire company), it might be far less expensive to buy that equipment rather than wait three days per sample.

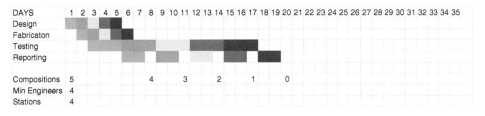

Figure 10.5 Staggered-serial approach to workflow management.

- Outsource the task. Could you pay someone else to run the testing as a service? It is good practice to replicate part of the testing internally to check the quality and repeatability of the outsourced effort.

Once the main bottleneck has been addressed, you can start to improve the remaining parts of the development project. The previous examples used a simple case where only five subsequent material synthesis iterations were needed in order to reach a functionally satisfying result for the company. In the case of newly developed materials for new applications, the number of iterations required will likely be orders of magnitude higher. If, for example, we have a total addressable experimental space of 1000 samples and each one takes one day, the timescale becomes prohibitive for any company, let alone a start-up. Optimization techniques such as DoE are somewhat effective in reducing the number of iterations required, while exploring the effect of interdependencies between process parameters. However, DoE can only reduce the total number of iterations by 30–50% at best and only if the governing process parameters have already been identified [9].

Large sets of unique materials are better addressed by state-of-the-art combinatorial material synthesis and, in some cases, analytical techniques. Combinatorial techniques are well-established and have been applied in catalyst screening, photovoltaics, data storage, and more. Its distinct advantage is the ability to synthesize and potentially characterize an almost infinite combination of materials in a short amount of time relative to a serial method.

Automated testing routines become vital now that thousands of material combinations are readily available for evaluation in a short amount of time. At the same time, the amount of data generated by the R&D team will suddenly increase 100–1000 times. Processing and storing this vast amount of valuable data in a structured and traceable way is as important as the process of gathering it in the first place. Conventional lab books and spreadsheets will not be effective at this scale. A well-designed and maintained structured database is the most sensible way to store the company's data, IP, and know-how. This database can also be easily analyzed with modern machine learning (ML) techniques to extract additional value from both positive and negative results. ML has been successfully applied in various fields within material science including high entropy metals, photovoltaics, data storage, and many others [10]. Figure 10.6 shows the high-level flow of activities during a combinatorial screening project.

10.9   Growing and Evolving the Team

Your company is rapidly growing from idea, to concept, to prototype, and finally to product. As we mentioned, the team has to grow with the company. The founders created a concept from an idea, the misfits (term of endearment) advanced it to a prototype, and now it is finally time to concentrate on creating the commercial product. What kind of team will be able to do that?

It is critical to hire a team only after the project has been defined and the key skills necessary to execute on it have been identified. The notion of defining a problem prior to hiring the people to solve it sounds rather obvious when in fact the opposite is often

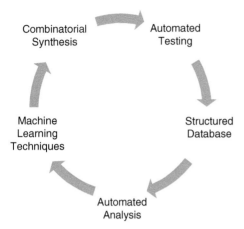

Figure 10.6 High-level flow of activities in a machine learning–assisted combinatorial screening project.

practiced. Decision-makers have a tendency to revert to what worked "the last time," meaning the Seed stage. An early stage start-up is often ripe with generalists: people who know a little bit of everything, are insanely quick learners, and don't mind working on different things every day. Great generalists are profoundly important at the beginning of a company, when no one knows what the first product will be or how it will be made.

Reaching the Series-A stage usually means that these uncertainties have been solved and a clear path toward commercialization has been chosen. Later-stage companies need the exact opposite of generalists; they need specialists who have the skills and the mentality required to execute on the commercialization plan in its entirety. These are often chemists or professional engineers with years of industrial experience in fast-changing and financially constrained environments.

10.10 Summary

This chapter presented an overview of our experience initiating, growing, and managing the technical program of a materials-based start-up company. The most important phases of early growth were presented and discussed with particular emphasis on setting appropriate technical objectives and building the right team for each phase. The most common development cycles were described with attention to identifying and overcoming bottlenecks. Finally, modern approaches to sustainable chemistry and materials R&D have been presented including the use of combinatorial synthesis, structured databases, automated data analysis, and the added value of machine learning. In our experience, most companies do not fail because of the inherent technology – they fail because of poor management. Bear this in mind while you use and adapt the skills and approaches detailed in this chapter. The investment of your time and effort in honing these skills will undoubtedly pay off in the long run.

References

1 Wuttig, M. and Yamada, N. (2007). *Phase-change materials for rewriteable data storage. Nature Materials* 6 (11): 824–832.

2 Lam, C.H. (2014). Phase change memory and its intended applications. *2014 IEEE International Electron Devices Meeting*, San Francisco, CA (15–17 December 2014), IEEE.

3 Kolobov, A.V. et al. (2004). *Understanding the phase-change mechanism of rewritable optical media. Nature Materials* 3 (10): 703–708.

4 Hosseini, P., Wright, C.D., and Bhaskaran, H. (2014). *An optoelectronic framework enabled by low-dimensional phase-change films. Nature* 511 (7508): 206.

5 Hosseini, P., and Bhaskaran, H. (2015). Colour performance and stack optimisation in phase change material based nano-displays. *Proceedings Volume 9520, Integrated Photonics: Materials, Devices, and Applications III SPIE Microtechnologies*, Barcelona, Spain. SPIE.

6 Anthony, S.D., et al. (2019). Breaking down the barriers to innovation. *Harvard Business Review* (November–December).

7 Miller, J.H. and Page, S.E. (2007). *Complex Adaptive Systems : An Introduction to Computational Models of Social Life*, Princeton Studies in Complexity, vol. xix, 263. Princeton, NJ: Princeton University Press.

8 Eppinger, S.D. and Browning, T.R. (2012). *Design Structure Matrix Methods and Applications*. Cambridge, MA: MIT Press.

9 Anderson, V.L. and McLean, R.A. *Design of Experiments : A Realistic Approach*, 11e. CRC Press.

10 Ramprasad, R. et al. (2017). *Machine learning in materials informatics: recent applications and prospects. npj Computational Materials* 3 (1): 54.

11. Bridging the Gap 2: From Validation to Pilot Scale-Up

11.1

Part 1: Setting the Groundwork

James Lockhart and Andrew Ellis

NORAM Engineering and BC Research, Vancouver, BC, Canada

11.1.1   Introduction

The previous chapter explored key concepts on how to advance your start-up company's innovative sustainable chemistry technology, from the discovery to validation phases. Now that the basic technology has been proven and more funding has been raised, it is time to take your product development to the next level. This involves increasing the size of your production process to begin demonstrating that product can be made in large quantities. Bench-scale setups simply will not suffice at this point. What is needed is a design for a large-scale production process that can be done economically and efficiently. This is done through process scale-up.

This chapter will focus on the progression from bench-scale experiments through the pilot-scale stage. Loosely defined, "pilot" scale is the first intermediary step between bench-scale and commercial manufacturing. Depending on the process, one pilot unit may be all that is required before full commercial scale. However, for larger and/or more costly ventures, additional stages may be required to de-risk the project. The last and most complete of these is often called a "commercial demonstration unit." This stage incorporates all of the key elements, but may be a somewhat smaller scale to reduce costs and/or not intended to be run for a long period of time.

Scaling up is necessarily a large topic and likely to be unfamiliar to the sustainable chemistry innovator. To make its introduction easier, the topic will be divided into two parts. This first part will focus on main cost drivers, risks, unknowns, and data required for successful scale-up. Understanding these factors is key as you begin the engineering groundwork to scale up your process.

The following are the topics covered in this part:

- **Letting go**: The independent-minded innovator must understand that at some point you must start to use and trust outside expertise.
- **Safety**: As things get bigger, so do the risks, and safety must be your top priority. Larger reactions, processes, and equipment are inherently more hazardous and must be effectively de-risked so that the health and safety of you and your workforce is protected.

How to Commercialize Chemical Technologies for a Sustainable Future, First Edition.
Edited by Timothy J. Clark and Andrew S. Pasternak.
© 2021 John Wiley & Sons Ltd. Published 2021 by John Wiley & Sons Ltd.

- **Commercial considerations**: This means understanding what the market actually wants and assigning its appropriate value. Commercial considerations are key to understanding and effectively planning the scale-up phase(s).
- **Technoeconomic analysis**: Here the engineering work begins and involves many new concepts that may be unfamiliar to the chemist or materials expert. It includes developing a mass and energy balance to understand all of the ins, outs, and intermediate flows within the process and getting a handle on process economics.

Part 2 will focus on the actual piloting and will include more advanced engineering concepts.

It is important to first realize that process scale-up is in itself a process. Rarely does a technology successfully go from a bench-scale process to a large manufacturing plant in a single step, though many have tried and failed. Like all the other aspects of commercialization described in this book, it requires multiple steps and adjustments along the way as new information is available or challenges arise that must be overcome. However, if the general procedures are understood ahead of time, you can minimize the costs and time required. Working with outside experts will be helpful as the skill sets required for scale-up are different from those required for the early development stages.

NORAM, and its wholly owned subsidiary BC Research, have been developing, scaling up, and commercializing new chemical processes and technologies for more than 30 years. During this time, we have seen numerous successes and failures and developed considerable experience in this sector. Although each technology's commercial path is unique, this chapter's intention is to provide some common keys to success and pitfalls to avoid that we have observed over the years.

11.1.2   Letting Go and Obtaining External Expertise

Start-ups in the CleanTech sector are often led by highly qualified experts in a particular scientific field such as chemistry. These champions are critical to the success of the venture; however, they cannot be experts in everything. It is necessary to assemble a diverse team skilled in other areas, particularly engineering, which includes process design, costing, specialized equipment design, and scale-up.

Editor's Note

This is often a challenge for the independent-minded entrepreneur, particularly those from the basic sciences. Bringing in others with differing expertise is a necessity, whether the individual is hired to be part of the company or as an external consultant.

Here is another truth. If you expect to raise any level of financing close to what is needed to ultimately get your product to market, you must bring in expertise beyond your capabilities. If you are not comfortable bringing in engineering expertise at this stage, it will be forced upon you by your investors. Investors are looking for management teams that are well rounded and have the required expertise beyond the initial innovation discovery.

Individuals and companies should recognize gaps in their expertise and be willing to work with others to avoid re-inventing the wheel or falling victim to blind spots

---

or problems that are well known by experts in that field. A great many start-ups have failed due to the unwillingness of the initial founders or inventors to "let go" and bring in additional expertise.

One question that often arises is how much expertise should be hired versus contracted out to experts and consultants? Although employees are less expensive per hour, they represent a much longer-term commitment that cannot be as quickly adapted in response to project demands and available financing. Outside experts can be brought in relatively quickly for smaller and more focused projects as required allowing a much greater breadth of experience without large ongoing costs. As with due diligence undertaken during the hiring process, verify the experience and track record of an external consultant before engaging them.

Protecting confidentiality and intellectual property is important (as detailed in Chapter 4) but needs to be balanced against the dangers of the team going it alone through unfamiliar waters. This often means engaging an intellectual property expert.

There is no choice but to work with others to help maximize your chances of success. If you can effectively cross this threshold, your company will substantially accelerate its growth and be poised for future success.

11.1.3   Safety Considerations

Safety is paramount and must be considered at all stages of development, but it is important to recognize that generally *hazards and their associated risks increase with scale*. Beyond the obvious use of larger volumes of chemicals, there are also greater amounts of energy stored in pressurized, flexible, and rotating/moving equipment. Additional factors such as lower relative rates of ventilation, weaker equipment (lower wall thickness to diameter ratios), and reduced natural cooling are just a few things to consider.

An important example of this last point are exothermic reactions. At bench scale, it is often easy to achieve high amounts of cooling due to the small volumes used, and in fact heating may even be required to maintain the desired temperature. As the scale increases, the same exothermic reaction will require far greater attention to ensure adequate cooling and temperature control. This is due to the reduced surface area to volume ratios and resulting reductions in relative thermal mass and natural cooling. Without careful consideration of the heat released by the reaction and design of the cooling system, inadequate cooling at larger capacity may result in higher than desired temperatures. This can increase reaction rates further, resulting in a "run-away" reaction that reaches very high temperatures with potentially catastrophic results.

To help identify hazards and mitigate risk, standardized process review methods have been developed. These are especially important when developing new processes that are less well understood. An example of one such method is a Hazard Identification (HAZID) study, typically done following the initial process design, but prior to the detailed design phase or any larger-scale testing. During a HAZID study, the design team reviews a series of key words or phrases that point to potential hazards for each unit operation or process area

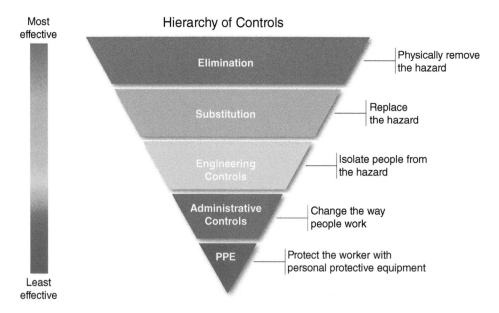

Figure 11.1.1 Hierarchy of controls. Source: From Wikipedia, Centers for Disease Control and Prevention.

and considers their applicability. In this way, a detailed list of the main risks to personnel, the environment, equipment, and company reputation are identified and compiled.

Once hazards are identified, the associated risks can be mitigated throughout the rest of the design using what is known as a "hierarchy of controls" (see Figure 11.1.1). These range from complete elimination of the hazard to the use of personal protective equipment (PPE).

Typically, the earlier a hazard is identified in the design process, the more likely an effective method of control can be implemented. This is because the most effective controls (e.g. elimination) typically change the process significantly, so it is more straightforward to implement them in an earlier phase.

For example, a change in catalyst or solvent may significantly reduce or eliminate a particular hazard. However, it may also result in a very different impurity profile and have significant impacts on the rest of the process, product quality, or emissions. If considered early in the design phase, this change may be manageable and even simplify the process. If left too late, it could necessitate major changes to processes that have already been developed in great detail. Later in the development there is greater pressure to apply engineering controls, administrative controls, and personal protective equipment to avoid additional costs and delays even if the more radical change represents the most effective solution. This is one reason not to rush the early development phase, skip important steps such as the HAZID, or hesitate to involve qualified experts.

As the design is further developed, more detailed process hazard analyses (PHAs) are performed. This includes "what if" analyses and hazard and operability studies (HAZOPs). These should be conducted by a diverse team of qualified personnel that include all relevant engineering disciplines, operations, and maintenance personnel. In the PHA, the design is tested by considering different potential operational upsets (e.g. high-pressure spike) or other conditions (e.g. unexpected shutdown) for individual pieces of equipment or systems. When potential hazards are identified, the risk is ranked based on probability/frequency and severity with and without existing controls and PPE. Where required, additional controls or changes are made to the design to ensure the potential risk is deemed acceptable by the proponent.

11.1.4 Commercial Considerations

A key determinant for the successful commercialization of a chemical technology is a solid understanding of the commercial product your company hopes to create. Many of these have been described in previous chapters but will be summarized here from a piloting perspective. The main goal is to understand exactly what you are going to sell and what the value of these products is to the market (which may be very different than what you think it should be).

11.1.4.1 The Market

As detailed in Chapter 2, when developing a new technology, it is critical to identify the target market(s) and its key requirements and drivers as soon as possible. It is quite common for the perspectives of innovators and entrepreneurs, particularly fresh from academic environments, to markedly differ from external consultants and actual plant owners/operators who are intimately involved in industrial operations. Markets are influenced significantly by nonintuitive factors, sometimes based on historical considerations. An ounce of experience at this stage can help avoid wasted time and effort addressing perceived problems that are not actually significant issues in the industry. Even worse than this is failing to address an unforeseen issue until later in the development or scale-up stages, requiring significant reconfiguration.

In the case of a new process to make an established commodity product (identical to that already on the market), adoption may be almost immediate. On the other side, it may take a decade or more for a completely new product such as aerogels or nanomaterials to be widely accepted and adopted. Even with expert help, it is quite common to underestimate the time required for a new technology to reach and become accepted by the market. This varies depending on how different the technology or product is compared to established ones, anticipated lifetime of the product (e.g. structural materials versus disposable single-use products), reasons for adoption (e.g. cost, performance, legislative necessity), and perceived risk.

11.1.4.2 Location Constraints and Opportunities

Although it is not necessary to have exact site locations identified early in the development process, it is helpful to understand the type of location and jurisdiction that are most likely as these can have a large impact on the scope and cost of the plant. In particular, it is important to understand the following:

- Availability and cost of energy, water, feed chemicals, and disposal facilities.
- Transportation and access of product to market that may impact its final form.

- If the process is to be incorporated within another plant, this will impact plant capacity, available footprint, automation requirements, as well as the battery limit conditions, applicable standards and other facility constraints, etc.
- Co-locating with other industries could offer reduced infrastructure costs, lower-cost utilities/waste energy utilization, by-product usage/recycling, waste handling, shared labor and other resources, etc.

11.1.4.3 Appropriate Feedstocks

At lab scale, most chemicals are available in a variety of concentrations and purities. Initial research and development (R&D) studies are typically performed using relatively high-purity chemicals to simplify experiments, results, and analysis. However, these chemicals may not be cost-effective or even available at the quantities required as the process is scaled up.

Material costs in small quantities are often dominated by labor, overhead, and profit, such that large differences in the underlying production costs are often obscured. It is essential to determine the true industrial cost and availability of the required feedstocks in the desired purity. This is difficult to do in many industries and can vary widely by location. Readily available information on the Internet for online purchase from overseas or historical spot prices is often not reflective of the true cost for a large-volume negotiated contract. Location can also have a major impact on shipping costs, the availability of different grades, as well as the underlying feedstock's production costs due to differences in labor and energy costs, for instance.

At industrial quantities there are often only a few grades of raw materials commonly available, and these typically contain levels of impurities many times higher than used at lab scale for R&D purposes. Some suppliers may be willing to provide specialty grades, particularly for use in higher-value/lower-volume sectors. However, this will add to the cost and reduce supply flexibility and competition. In addition to the amount of impurities present, the nature of the impurities may be different due to variances in process or purification techniques at different scales.

It is important for the development team to work with suppliers to learn the nature of the impurities present and test representative feedstocks to understand how these impurities may affect their process, as it can have major implications on capital and operating costs. In an ideal case, differences in feedstock purity may only affect final product quality. In reality, they can also impact reaction rates and selectivity, corrosion rates, and the lifetime of catalysts or other critical components, especially if they tend to accumulate due to process recycles. To address these issues, additional unit operations may be required including upstream purification, downstream separation, or waste treatment and disposal, all of which impact plant capital costs and overall economics.

11.1.4.4 Co-product Value

Co-product sales can greatly improve the process economics by turning a potential liability and cost into a revenue stream. However, caution is required to avoid over-estimating the net benefit.

First, it is important to consider the additional capital and operating costs required to separate, purify, beneficiate, and sell these co-products. These may be substantial, especially if trying to achieve a standard commodity-grade from a nonstandard process. Production of nonstandard grade products may be significantly less expensive to achieve but will be much more difficult to find buyers for, especially at small volumes. It also typically requires significant discounts from the analogous commercial-grade commodities.

Establishing strategic partnerships, which take advantage of synergistic co-product situations such as location and/or impurity profiles, may be possible. However, they require significant effort and are by no means assured.

11.1.4.5 Downstream Product Upgrading

A common mistake made when the target process does not demonstrate the anticipated beneficial economics is to tack on additional, often conventional, downstream processing to produce a higher-value product. This is not a valid approach if existing technologies can generate the initial product more economically than the new process. A new process needs to stand alone on its own merits. Further, it must demonstrate significant benefits over the incumbent technologies to be successful given the additional risks and costs involved in its implementation and commercial deployment.

11.1.5 Techno-Economic Assessment

A key step at this stage is the techno-economic assessment. This involves creating a methodology to analyze the technical and economic performance of a process, product, or service. In the case of a sustainable chemistry or materials product, the first steps are to create the mass and energy balances. These involve understanding and quantifying the materials and energy involved in the larger-scale process so that costs can be attributed to them. This in turn allows the sizing of the various unit operations and estimation of capital and operating costs of the process at larger scales.

11.1.5.1 Case Study Example

We will introduce a case study that will carry through this section. Suppose a new technology is proposed for production of a high-value food additive from a low-value food processing by-product in aqueous form. The liquid is oxidized at high temperature and pressure to yield 5 wt% of the product. This is recovered from the reaction mixture through solvent extraction; then the solvent is removed and recycled, leaving the product in solid form.

11.1.5.2 Mass Balance

A mass balance is performed to obtain an understanding of the major material flow rates into, out of, and within the process. It is based on a series of calculations that track the various masses of materials through the process. If you draw a box around an individual unit or group of units, mass that goes in must equal mass that comes out. Materials cannot

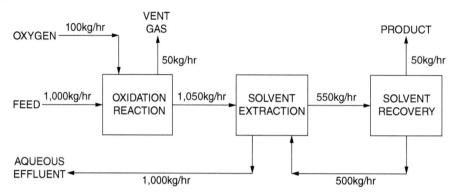

Figure 11.1.2   Example block flow diagram.

disappear or accumulate when the process is at steady state (in = out at steady state). This is the essence of the mass balance.

Different degrees of detail are typically modeled in the mass balance at different stages in the development process.

- During initial evaluations, major unit operations involving transformations and separations are modeled as single "blocks" and illustrated in a block flow diagram (see Figure 11.1.2). At this stage, each block may represent many different pieces of equipment, and even completely different operations for various design cases. Only the process flows (often called "streams") in and out of these blocks are considered, not those happening within the blocks themselves.
- As the design progresses, additional details within each block are considered, and unit operations and equipment types are selected. The chemical flow rates and compositions between these are determined and illustrated in one or more process flow diagrams (PFDs). Figure 11.1.3 shows an example using the oxidation reaction. For the purpose of this high-level case study, we are not going to worry about what the different symbols mean.

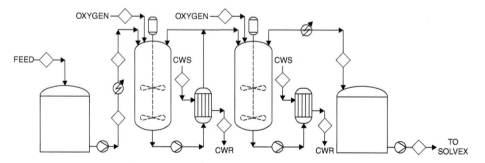

Figure 11.1.3   Example process flow diagram (PFD).

- Eventually the entire process is modeled including all the individual pieces of equipment and the process flows between them. In addition to the chemical (component) mass flows of each stream, the model may also simulate and/or summarize temperatures, pressures, and fluid properties.

When starting with a new process, it is inevitable that there will be many questions and unknowns. These should not be a deterrent to performing the mass balance – by proceeding on reasonable assumptions, the mass balance can help you identify gaps in the available information.

The relative significance of these technical and commercial unknowns and required assumptions can be explored through a sensitivity analysis (discussed later in Part 1). Creation of a mass balance will force the designer to determine the following:

- Flow rates and compositions of the flows leaving the process
- Reactions that are occurring and any by-products or co-products that may be formed (reaction yields are rarely 100%)
- The fate of various by-products and feed impurities (how they separate across each unit operation)
- The impact of by-products and impurities on each unit operation and what separations may be required to control by-products and impurities to acceptable levels
- Potential actions to ensure a safe, cost-effective and environmentally acceptable process

Mass balances can most easily be prepared using process simulation software. Unfortunately, the initial results from these process simulators are often inaccurate as they only typically estimate process parameters and chemical interactions, especially in concentrated aqueous and other nonideal systems. Unless a designer is diligent in checking the program's underlying calculations and assumptions and the veracity of the results, critical unknowns may go unnoticed and untested. This is especially challenging when dealing with chemical species not built into the process simulation software. Manual spreadsheet-based mass balances can also be used, which force the designer to consider what is going on at each stage of the process. However, these are more time-consuming, require a high level of expertise, and have a much greater risk for arithmetic errors.

In either case, a mass balance needs to undergo rigorous checking and review both to ensure mathematical accuracy based on the assumptions and intended methods and to experimentally validate the resulting predictions.

11.1.5.3 Energy Balance

Next up is the energy balance. It's similar in concept to the mass balance, except that instead of calculating the mass in and out of individual unit(s), energy is analyzed going in and out of each operation. As with the mass balance, a box can be drawn around a unit operation, or multiple operations and the amount of energy going in and out in a steady state system must be equal.

For chemical processes, energy calculations are based on enthalpy – a measure of the total heat content of a system or the amount of energy for each mass or mole unit of material (e.g.

kJ/kg or kJ/kmol). Enthalpies of each stream are calculated based on the sum of the heats of formation, plus adjustments for temperature based on heat capacity and phase change as appropriate. Theoretically, enthalpies of mixing are also required but are typically fairly modest and thus can often be ignored except for very nonideal mixtures (e.g. strong acids and bases).

For many common fluids, data and correlations for enthalpies of formation, phase change, and heat capacity are readily available from many sources (e.g. *Perry's Chemical Engineer's Handbook*, International Critical Tables). In the case of less common chemicals where data is not easily obtained, it is possible to estimate using standardized methods such as the Joback method, which is based on molecular structure.

Early in process development, a complete energy balance is typically not required. Except for processes with exceptionally large energy demands or costs, the main energy demand and costs can be estimated and accounted for on an as-required basis. However, as the design is refined, individual energy demands become important for equipment design, and a detailed energy balance becomes necessary.

11.1.5.4 Process Economics

Once the basic mass and energy balances have been carried out, the next step is to use this information to determine the costs associated with your scaled-up process. These costs are generally divided into two basic categories:

- **Capital costs**: This is the cost associated with initial purchase, fabrication, and installation of equipment and the fixed physical aspects of your process. Reactor vessels, piping, and system control equipment all fall into this category.
- **Operating costs**: This includes the costs of running and operating the system. Raw materials, utilities (electricity and water), labor, maintenance, and waste treatment are all considered in this category.

Capital cost analysis is typically done in several stages. Accuracy increases as the design progresses from an early-stage order of magnitude estimate (based on similar systems) through to detailed estimates with an accuracy of ±10% or less. The latter can be achieved once detailed design has progressed significantly and thorough quotes for all of the equipment and major contractor packages are obtained.

There are several different industrial conventions for defining the scope and anticipated accuracy of capital cost estimates. A common convention is the "class system," which categorizes Class 5 (least accurate) to Class 1 (most accurate). Although not as accurate, the lower-effort, earlier class estimates are still of value in understanding the approximate costs involved and helpful in evaluating the cost–benefit of major process and equipment changes. The American Association of Cost Engineers (AACEs) has defined standardized cost estimates to help bring consistency and better accuracy to the process industries.[1] The following table summarizes the key features of the various classes of estimate:

---

1 AACE International. (2011). Recommended Practice No. 18R-97 COST ESTIMATE CLASSIFICATION SYSTEM – AS APPLIED IN ENGINEERING, PROCUREMENT, AND CONSTRUCTION FOR THE PROCESS INDUSTRIES TCM Framework: 7.3 – Cost Estimating and Budgeting Rev. AACE (29 November).

Estimate class	Maturity level of project	End usage	Methodology	Expected accuracy
	(expressed as % of complete definition)	(Typical purpose of estimate)	(Typical estimating method)	(Typical variation in low and high ranges)[a]
Class 5	0% to 2%	Concept screening	Capacity factored, parametric models, judgment, or analogy	L: −20% to −50% H: +30% to +100%
Class 4	1% to 15%	Study or feasibility	Equipment factored or parametric models	L: −15% to −30% H: +20% to +50%
Class 3	10% to 40%	Budget authorization or control	Semi-detailed unit costs with assembly level line items	L: −10% to −20% H: +10% to +30%
Class 2	30% to 75%	Control or bid/tender	Detailed unit cost with forced detailed take-off	L: −5% to −15% H: +5% to +20%
Class 1	65% to 100%	Check estimate or bid/tender	Detailed unit cost with detailed take-off	L: −3% to −10% H: +3% to +15%

a) The state of process technology, availability of applicable reference cost data, and many other risks affect the range markedly. The value represents typical percentage variation of actual costs from the cost estimate after application of contingency (typically at a 50% level of confidence) for a given scope.

In early-stage estimates (e.g. Class 5 and possibly Class 4 under some circumstances), it is feasible to link a capital cost estimate directly to the key outputs from the mass and energy balance. A base equipment cost is selected and then adjusted (scaled) based on key sizing criteria (volume, weight, surface area, throughput, etc.) depending on the type of equipment, unit operation, or even entire plant. This adjustment typically takes the form of a $Cost_2 = Cost_1 * (Size_2/ Size_1)^n$ where n = 0.4 to ~1.0, and most commonly ~0.6–0.7. Thus, many pieces of equipment that double in size only increase in cost by $2^{0.65} = $ ~1.5x (a ~50% increase).

An exponent below 1.0 means that the item of interest becomes less expensive as it increases in size or capacity – the economy of scale. Base unit costs, capacity, and exponents may be found in various references[2] or calculated using multiple quotations for similar pieces of equipment. Equipment costs can also be adjusted for secondary factors such as pressure, temperature, and material of construction.

Equipment cost adjustments should also be made for currency fluctuations and time, often using an appropriate price index (e.g. Chemical Engineering Plant Cost Index[3]) that accounts for inflation as well as changes in underlying material costs, and other factors for similar equipment over time.

---

2 Perry, R., and Green, D. (1985). TABLE 9-50 Typical Exponents for Equipment Cost versus Capacity. In *Perry's Chemical Engineer's Handbook Version 6*. McGraw-Hill.
3 The Chemical Engineering Plant Cost Index (CPI) has been published monthly by *Chemical Engineering* magazine since 1963. There are other cost indexes published by others as well for different types of equipment.

Even with all of these adjustments, the accuracy of scaled equipment or plant costs are not very high. This is due to changes including available equipment sizes, quantities/methods of production, labor for design customization, changing profit margins, differences in scope of supply, and warranties, and, as a result, scaled costs should be used only for early-stage estimates. Where greater accuracy is required and/or limited cost and scaling data is available, supplier quotations should be obtained.

In the most accurate, detailed cost estimate, all the required components and installation items will be itemized and costed. However, in an early-stage cost estimate, equipment costs are often multiplied by installation factors to arrive at a total installed cost. Installation factors range from ~1.5 - 2.5x for pre-piped and instrumented skid-mounted units to ~5x for piecemeal tanks, and pumps. This typically results in an average installation factor of 3–4x for the entire plant.[4]

For example, at one extreme a fully assembled, pre-piped and instrumented water purification skid might cost $500 000 to purchase. The total cost required to finish the design and installation of that unit operation is 1.5–2x ($750 000–$1000000) for the site preparation, concrete pad, power, piping to/from the system, building around it, and controls integration.

In contrast, a heat recovery system consisting of several tanks, pumps, and heat exchangers also costing $500 000 to purchase might have a total installed cost closer to 4 or 5x the equipment cost ($2 000 000–$2,500 000) due to the greater amount of engineering, equipment installation, instrumentation and electrical, piping and valves, insulation, and foundations that are required, but not already part of the equipment cost.

Although some installation factors may be found in the literature, they are more accurate when based on experience with actual project costs involving similar types of equipment, capacity, and industry (which often have substantially different requirements for automation, materials of construction, quality/standards, reliability, etc.).

If the capital cost estimate is linked with the mass and energy balances and adjusted automatically as described, you can evaluate the impact of changing those balances on capital cost based on how multiple unit operations are affected. Given the relatively low degree of accuracy of this methodology, only large differences will be definitive. Nevertheless, it can still be a valuable tool to help identify major differences, trends, and options that are worth investigating in subsequent development stages. For example, suppose a new catalyst has been considered for implementation in your process. It is highly corrosive, and so will require high alloy metallurgy throughout the reaction section, but it triples the yield of the process, so some of the equipment becomes correspondingly smaller. Both cases can be assessed with a Class 5 estimate to help indicate if the move to the new catalyst may be justified.

Once capital costs are estimated, outputs from the mass and energy balance can be extended to begin estimating the operating costs. This includes analyzing feedstocks, utilities/energy, and effluent handling to arrive at the key variable operating costs and margins. In addition to the variable costs described, it will eventually be important to consider more fixed costs such as property taxes, insurance, capital repayment and interest, labor, and maintenance.

---

4 Perry, R., and Green, D. (1985). TABLE 9-51 Factors to Convert Delivered-Equipment Costs into Fixed-Capital Investment. In: *Perry's Chemical Engineer's Handbook Version 6*, McGraw-Hill.

It is useful to perform one or more separate analyses of the process economics under different operating conditions. It is far more powerful if these can be linked to the mass and energy balances, which will allow the analysis to scale reasonably to any changes in the underlying process. This type of live economic model is better able to assess the net impact of different, and sometimes simultaneous, changes in the process that are often not obvious from inspection alone.

11.1.5.5 Sensitivity Analysis

Preparation of the mass and energy balances typically results in an intimidating list of unknowns and required assumptions. It is unlikely that there will be sufficient time or resources to address all of them in depth, so a way of identifying those that will be of the most material significance to the technical and commercial success of the technology is needed. This method is known as "sensitivity analysis," a powerful tool to identify the most important inputs and assumptions on which to focus the technology development efforts.

In the simplest form, sensitivity analysis is performed by increasing and decreasing any assumed value and observing the impact on the key outputs from the mass and energy balance. Often there are offsetting impacts in different areas of the process that are difficult to evaluate by visual inspection, so it is helpful to link the mass and energy balances to operating and capital cost models so that the overall impact can be quickly observed. A sensitivity analysis is typically performed on key inputs individually; however, with experience, many can typically be dismissed as less important. Others deserve more detailed inspection, often including perturbing multiple inputs simultaneously to study interactions.

For example, if a solvent is to be distilled from your product and its heat of vaporization is unknown, the energy input requirement can be assessed using an estimated value. Sensitivity analysis can then be performed by assessing the energy input requirement at a high bound and a low bound estimate. If the calculated range of energy requirements leads to a major operating or capital expense relative to other costs in the process, it will be important to conduct further trials to ensure the heat of vaporization is known with accuracy. If it is not significant, it may be appropriate to use an estimated value without further validation until later in the design phase.

11.1.6 Conclusion

The intention of this first part of this chapter was to give you an overview of initial concepts related to the planning and design aspects of scale-up and piloting. Although some of the language and challenges related to scale-up and piloting may be new to you, a strong professional team with experience in new technology development will be able to help you prepare a comprehensive development plan and assist you through each step to ultimately bring your commercial product to market.

At this point, the actual physical engineering work involved in the scale-up process can now begin. This is covered in the second part of this chapter.

11. Bridging the Gap 2: From Validation to Pilot Scale-Up

11.2

Part 2: Building the Pilot Unit

James Lockhart and Andrew Ellis

NORAM Engineering and BC Research, Vancouver, BC, Canada

11.2.1   Introduction

This part moves into more advanced engineering and mathematical aspects of the scale-up process, as well as the actual physical construction of a pilot unit. Although more detailed technically, it is still only an overview to introduce the innovator to key concepts. These are the topics covered in this part of the chapter:

- **Piloting and scale-up basics**: The basic concepts are introduced including motivation and objectives.
- **Process and equipment considerations**: There are many process and equipment design choices to be made at the pilot scale in terms of cost, effectiveness, and time.
- **Pilot plant location and operation**: Where you install the plant and how it operates can have a dramatic effect on its ultimate cost and profitability.

Note that many of the topics covered here contain mathematical and theoretical concepts that assume a basic knowledge of chemical engineering. Don't be too concerned if the details are not immediately grasped. The purpose is only to give an overview so that you can better understand the scale-up process, assess/select your scale-up partners, and communicate effectively with them during the process.

11.2.2   Piloting and Scale-Up Basics

11.2.2.1   Motivation

Having considered the commercial goal for your sustainable chemical technology and determined the main cost drivers, risks, unknowns, and data required for successful scale-up, the next step is to test at the pilot scale. This is not the end goal, but a tool to assist with technology development by simulating the key elements and unit operations to be used at larger manufacturing scales.

There is often a desire to build as big a pilot as possible as soon as possible by scaling up the exact types of equipment used in the bench scale. However, this is often not the right approach because of the following:

How to Commercialize Chemical Technologies for a Sustainable Future, First Edition.
Edited by Timothy J. Clark and Andrew S. Pasternak.

- Changes in process or equipment that may be required at the future commercial scale
- Design data requirements for scale-up
- Constraints such as available feedstocks, utilities, space, or most importantly budget and schedule

In some cases, there may be pressure to skip a pilot altogether. This may be tempting if you have done a considerable amount of testing at bench scale or feel so confident that you just want to make it happen quickly (the "Hail Mary" approach). Even when there are no major questions identified that need to be answered, a pilot is still helpful to reduce risk (both real and perceived) at the subsequent scales. It may identify and resolve previously unknown operating issues that arise with an increase in scale or when recycles are introduced.

A pilot should not be confused with commercial "demonstration" scale, which is typically at, or fairly close to, commercial capacity and incorporates most (if not all) of its elements. Commercial demonstrations are also typically designed for a longer life span than a pilot, though not as long as a commercial plant, with projected economics that approach break-even so that it can be run for an extended period of time (one to five years). In contrast, a pilot is an investment designed to advance the technology and de-risk the subsequent commercial demonstration and typically run for a few months to two years at most depending on the nature of the process and equipment.

Before beginning on the design or costing of a pilot, its objectives need to be carefully considered and clearly defined. These may include the following:

- Gathering design data for scale-up and addressing gaps in knowledge
- Gathering longer-term performance data for sensitive and critical elements such as catalysts, membranes, electrodes, etc.
- Identifying new unknowns and issues, especially in relation to impurities and by-products that accumulate in recycles
- Generating product in sufficient quantities for customer trials and testing
- Generating sample intermediate products for different suppliers
- Evaluating equipment for operational parameters and performance guarantee purposes
- Creating a technology showcase facility for potential customers and investors

Each of these objectives will influence the appropriate size (capacity) and costing of the pilot.

In addition, external constraints can also be imposed upon design and creation of the pilot unit. Most obvious are budget and scheduling issues. Availability of feedstock and equipment can also be challenging. Another often unforeseen constraint is where to actually place the pilot unit. Factors such as material handling, storage requirements, permitting, safety codes, and available utilities must all be considered. Experience and caution are required to avoid imposing a set of impossible constraints that ultimately lead to the failure of a pilot program.

11.2.2.2 Understand Critical Dimensions and Dimensionless Groups

Rarely is a successful scale-up achieved by matching only velocity or residence time. Rather, it is important to consider a number of critical dimensions and dimensionless groups depending on the situation.

Dimensionless groups are sometimes seen as a daunting, esoteric concept to those without engineering training. In reality, they are a simple and powerful means of characterizing the behavior of systems independent of physical scale, which provides a firmer basis for scale-up. There are many permutations of physical properties that yield dimensionless groups, but of most concern to process scale-up are typically those that pertain to macro-scale transport phenomena (e.g. fluid flow, droplet formation, heat and mass transfer, etc.).

The classic example familiar to most chemical engineers is the use of Reynolds number to characterize fluid flow behavior in different situations. Physically, Reynolds number describes the ratio of inertial to viscous forces over a relevant length scale. It can be applied to many fluid phenomena. As a simple example, consider a pipeline used to mix two immiscible fluids in a solvent extraction process. It has been shown that Reynolds number effectively predicts the onset velocity of turbulent flow across a wide range of pipe sizes, fluid densities, and viscosities.

The key point is that when designing the process, the engineer should ensure that the pilot reactor operates with a similar Reynolds number as the successful bench-scale experiments. Otherwise, process behavior may be markedly different, especially if flow crosses into a different regime (laminar versus turbulent).

Other dimensionless numbers effectively characterize other important and industrially relevant behaviors, such as mass transfer, heat transfer, mixing, and bubble/droplet formation. The following table lists some of the more valuable (and thus popular) dimensionless numbers used in scale-up. Similar to the earlier Reynolds number example, it is important that the pilot unit operate with dimensionless numbers similar to those in the smaller scale experiments.

Some common dimensionless numbers

Dimensionless number	Ratio	Application
Archimedes, Ar	$\dfrac{\text{Inertial forces} \times \text{buoyancy forces}}{(\text{Viscous forces})^2}$	Fluidized bed, spouted bed, and bubble column design
Biott, Bi	$\dfrac{\text{External heat transfer rate}}{\text{Internal heat transfer rate}}$	Heat transfer from sphere, e.g. cooling droplet
Fanning Friction Factor, f	$\dfrac{\text{Wall shear stress}}{\text{Velocity head}}$	Fluid friction in pipes, e.g. pressure drop.
Grashof, Gr	$\dfrac{\text{Buoyancy forces}}{\text{Viscous forces}}$	Heat transfer, e.g. natural convection
Froude, Fr	$\dfrac{\text{Inertial force}}{\text{Gravity force}}$	Fluid dynamics such as self-venting flow, ship hydrodynamics
Nusselt, Nu	$\dfrac{\text{Convective heat transfer rate}}{\text{Conductive heat transfer rate}}$	Heat transfer across a boundary, e.g. heat exchanger
Ohnesorge	$\dfrac{\text{Viscous force}}{(\text{Inertial force} \times \text{surface tension force})^{1/2}}$	(We/Re) Atomization, e.g. spray nozzles

Dimensionless number	Ratio	Application
Peclet, Pe	$\dfrac{\text{Convective transport}}{\text{Diffusive transport}}$	Heat, mass transfer, mixing
Prandtl, Pr	$\dfrac{\text{Momentum diffusivity}}{\text{Thermal diffusivity}}$	Heat transfer, e.g. relative conductive and convective heat transfer, momentum/thermal boundary layers
Reynolds, Re	$\dfrac{\text{Inertial force}}{\text{Viscous force}}$	Flow characterization (laminar versus turbulent transition)
Schmidt, Sc	$\dfrac{\text{Viscous diffusion rate}}{\text{Molecular diffusion rate (mass basis)}}$	Mass transfer analogy of Prandtl
Sherwood, Sh	$\dfrac{\text{Convective mass transfer rate}}{\text{Diffusion rate}}$	Mass transfer analogy of Nusselt
Weber, We	$\dfrac{\text{Inertial force}}{\text{Surface tension force}}$	Multiphase flows, e.g. bubble, drop formation

There are always trade-offs in scaling equipment and processes, and it is often not possible, feasible, or cost-effective to maintain full dimensionless group similitude for all aspects of a design. For example, physical limitations such as minimum available velocities to prevent solids settling, and maximum velocities to limit pressure drop, have to be respected even if they mean dimensionless numbers cannot be held constant. Often experience-based simplifications coupled with validation testing with physical models or simulations may be required.

In other cases, key physical dimensions may govern for practical or process-related reasons. In such cases, the commercial-scale performance may be better tested, and scale-up risk reduced, by using one or more elements matching those critical dimensions and scaling the other dimension(s), or number of units in parallel. For example, a tubular filter might be better tested by scaling the number of tubes used and their length rather than scaling their diameter and/or pore size.

The best methodology to test and scale each piece of equipment and system must be considered carefully, based on appropriate fundamentals and preferably using experience drawn from other successful designs and scale-ups.

11.2.2.3 Strategies for Success

It is rare that sufficient funds will be available initially to include everything you want in a pilot development program, so it is important to plan and budget well, prioritizing the most important elements from a cost and risk basis to reach the next important milestone. Strategies to reduce initial costs and increase cost-efficiency at this stage include the following:

- Building a pilot in stages to minimize up-front costs while demonstrating progression toward commercial deployment, potentially gaining support and funding for the next stage.

- Trying to leverage supplier testing facilities and equipment. They may be more willing than you think to help out if they see a potential benefit of the product or future sales in the long run.
- Using rented equipment to screen different technologies for the application before making final equipment selections. Delaying fixed costs until you are sure is always a sound strategy.
- Procuring used and refurbished equipment. No need to spend top dollar on new equipment if older equipment will suffice. This will not typically reduce installation costs but may improve the schedule as used equipment is often available much more quickly.
- Considering whether each unit operation must be tested. In some cases, there may be some conventional operations that are considered low risk, even at the future commercial scale.
- Only testing critical unit operations at the scale necessary to validate performance and/or feed subsequent steps. It is not always necessary for different parts of the pilot to be the same capacity.

Many of these concepts are discussed in further detail in the following sections.

11.2.2.4 Available Data and Its Limitations

Avoid investing time and money answering questions that have been addressed by others. This is incredibly important. Performing a thorough review of academic literature, patents, conference proceedings, and previous in-house research is a must-do activity. Even if the exact testing has not been performed before, similar work can save a huge amount of time and money. Be wary, though. It is sadly common for erroneous results to be reported in both the patent and peer-reviewed literature, so even where data sets appear complete, it is valuable to repeat experiments to validate results and gain a better understanding of the methods and systems employed.

While well-established and accurate methods have been developed for many systems, it is important to recognize that these may not give accurate results when applied to new systems. When developing analytical techniques for new and challenging systems, it is often best to use more than one method to try to confirm an important result. A powerful technique to help validate experimental results is to measure all of the inputs and outputs such that a mass, or preferably elemental balance can be prepared. When done properly, this can help to identify analytical inaccuracies and instill confidence that all significant species are being identified and quantified accurately.

11.2.2.5 Problems Specific to Smaller Scale

One of the goals of testing at small scale is to learn more about the process as well as identifying and resolving issues. It is often safer, faster, more flexible, and less expensive. Unfortunately, there are some additional challenges that arise at smaller scales that can hamper the effort and may not even apply at larger scales.

For example, smaller-scale reactions have higher surface area to volume ratios. This can result in increased relative heat losses that will introduce greater temperature changes and maldistribution without careful insulation and/or temperature control. In addition, higher percentage of flow near the reactor wall versus the bulk mixture impacts separations such as distillations, stripping, and scrubbing columns as well as mechanical equipment efficiencies (e.g. pumps and turbines). The greater relative contact with piping and equipment walls may increase pressure drops, contamination by corrosion, or promote side reactions.

One must be careful during small-scale experiments not to spend undue resources on parameters that do not represent larger-scale operations. Developing methods and strategies to avoid and account for these is important to maximize time and budget on studying issues most relevant to commercial scale.

11.2.3 Process and Equipment Considerations

11.2.3.1 Pilot Capacity

The desired capacity affects the cost and schedule of every other step in the piloting process. Once a capacity has been selected and detailed engineering work has begun, it can be difficult to change tack as many of the challenges and solutions are particular to the scale of the equipment involved. Plant capacity therefore warrants careful consideration early in the project.

Capacity should be assessed from a number of directions.

- **Commercial scale goal**: What is the proposed commercial capacity? How many stages of piloting/demonstration are planned? Working back from this, determine what capacity of pilot would be required to give technical confidence in the system at commercial scale. This sets a lower bound for the capacity.
- **Current scale of testing**: Evaluate the largest scale tests that have been performed to date. Based on this, think about the largest scale that can be designed with confidence. This sets an upper bound to the capacity.
- **Equipment availability**: Look at critical pieces of equipment projected for the commercial plant and identify the smallest models of the same types of devices that are readily available. This could define a minimum capacity for some key pieces of equipment if it is considered necessary to test it in the application prior to commercial deployment. Keep in mind that it is not always necessary for all parts of a pilot to operate at the same capacity, and intermediate tankage, slip streams, and/or different numbers of units may be a more cost-effective and technically adequate solution.
- **Market/testing requirements**: If the process generates a novel product, how much is required to meet existing and projected market demand? Are there specific quantities required for testing or customer samples? This could define a minimum overall plant capacity.

If the minimum scale identified is incompatible with the maximum scale for confidence based on current testing, additional stages of development may be needed. This could be achieved through bench testing, vendor testing, or potentially an additional stage of end-to-end piloting/demonstration. Where multiple steps are required, it is usually best to take the larger steps earlier where costs and hazards are reduced.

11.2.3.2   Batch and Continuous Operations

Chemical manufacturing operations can be generally divided into two main categories: batch and continuous. When developing the initial chemistry or proof-of-concept, batch processes are almost exclusively used. In a batch process, all the steps of the reaction are done in individual vessels, using a defined series of steps that results in the product. Raw materials are placed into the vessel at specific times; the reactions occur, product is removed, the vessel cleaned, and the process begins again for the new batch. Thus, batch operations have several separate stages of operation in a cycle that then repeat. These often include vessel fills with one or more feeds, heating/cooling to a desired temperature, completing a reaction, and finally separating via physical or chemical means (e.g. crystallization, distillation, filtration, or extraction/drying).

In contrast, a continuous process operates at the same conditions all the time (excluding start-up and shutdown), with raw materials fed into the front end of the process and product removed from the back end all on an ongoing basis. The reaction is not stopped at any time but allowed to proceed with raw materials constantly converted into product.

There are hybrids between batch and continuous processes. For example, a variant of the batch process is known as "semi-batch" operation, which has one or more species being introduced or removed continuously to all or part of an otherwise batch operation.

11.2.3.3   Batch vs. Continuous: Pros and Cons

Batch processes often require larger individual pieces of equipment due to the time required for feed and withdrawal as well as other start-up and shutdown considerations such as purging, heat-up, cool-down, etc. There are typically fewer, but more complicated, pieces of equipment as the same functionality is spread over fewer units. For example, one batch vessel may be used for multiple steps of a process such as, initial fill, heat-up, reaction(s), cool-down, settling, and emptying. Batch operations are typically flexible, which is one of their main benefits, especially for new processes where the optimal operating conditions may be less well known or are changing.

Alternatively, a continuous operation is often simpler and frequently lower cost overall. However, it may be more difficult to change their operating conditions, which are often linked (e.g. velocity/mixing and residence time).

All of these factors, and more, must be considered when deciding on the type of operation to be used in a given process. As with equipment type, it is important to understand whether batch or continuous operations are to be used at commercial scale and to ensure that testing at smaller scales will be representative. It is possible to use both types within a given process, but care is required to ensure adequate peak capacity and sufficient storage to buffer between batches.

The choice of batch versus continuous is often industry specific and ultimately depends on the economics of the process and regulatory requirements. For example, the pharmaceutical industry relies almost exclusively on batch processing in part because the final product is of extremely high value and made on relatively low (kg) scales with many carefully controlled synthetic reactions done in serial. In addition, individual batches can be more easily sampled, analyzed, and tracked, which is helpful for an industry with significant regulatory requirements and strict quality assurance/quality controls (QA/QCs).

At the opposite end of the spectrum is the petroleum industry, which produces titanic scales of relatively lower-cost products such as fuels, lubricants, and polymer feedstocks. In this case, economics demand a continuous process, which is seen in their gigantic refineries. In general, sustainable chemical technologies fall between these extremes, and whether continuous operations are preferred over batch ones will depend on the particulars of that process and the desired capacity.

11.2.3.4 Equipment Selection

At bench scale, most equipment is designed for versatility rather than minimum cost, long-term reliability, or efficiency. It is extremely rare that the optimal commercial design will simply be a much larger version of the early-stage bench-scale experiments. When scaling up to commercial production, the innovator must consider new factors, and acquaint themselves with a new lineup of available equipment types that are designed for the application at the target scale.

Although construction of a commercial unit may be in the distant future, it is important to identify the most likely unit operations and types of equipment to ensure the design and development program are appropriate to answer the relevant design questions. Much time and money may be wasted testing a particular operation at bench and pilot scale that is not available or suitable at commercial scale.

When dealing with novel and innovative processes, there may not be industrial designs available for the exact application at the scale required. However, by considering analogous materials and unit operations, commercial technologies used in other industries or processes may be identified and adapted. This offers significant benefits over starting from scratch and "reinventing the wheel."

Standardized designs and methods that have already been developed generally offer substantially lower risk and less expensive options than custom prototypes. The engineering work that goes into them has already been paid off, and there may be competition in the marketplace to keep prices down. Established technology/equipment suppliers may also be interested in partnering and sharing costs to develop a new potential product and market. Naturally, there will be some new risks involved with differences in application and associated system modifications. However, these are significantly reduced compared to a new, unproven technology or equipment type as many issues around equipment specific design, scale-up, fabrication, operation, and reliability will have already been identified and resolved. Furthermore, the technology will be available with much shorter lead times, and there are typically multiple capacities and other options available. Suppliers may have existing test units that can be purchased or rented and trialed in the new application.

Beware of step changes in equipment with respect to capacity. Some technologies have physical bounds to their ability to increase in processing capacity. Above that point, multiple units operating in parallel are required to achieve a desired capacity. This adds significantly to the installation costs compared to a single large unit. In some cases, multiple parallel systems may be unavoidable or even beneficial. However, this should be a consequence of evaluation, not a result of poor planning or foresight. Focus your development efforts on validating equipment and technologies that will best meet the needs of the future commercial scale.

11.2.3.5   Use Proven Equipment

Process development is challenging and has many unavoidable risks. To improve the odds of success, it is important to take on as few unnecessary risks and reduce using unproven operations as much as possible. This often runs counter to the nature of innovators who tend to gravitate toward the newest and most promising technologies. However, each new and unproven operation in a given process adds uncertainty and risk, jeopardizing the success of the entire business.

Innovators and entrepreneurs should work to maximize the incorporation of "off-the-shelf" equipment and processes, which are already proven and industrially practiced with minimal modification required. This requires stepping back from what may be perceived as an optimal design from an efficiency or cost perspective to maximize the probability of success for the core technology. In instances where it may be advantageous or synergistic to incorporate additional unproven technologies, it is valuable to consider alternatives such as smaller-scale side demonstration units or phased build-outs.

11.2.3.6   Suppliers/Sourcing

Once viable commercial equipment types have been identified, the next step is to find potential suppliers or vendors. Depending on the level of development, suppliers can be approached to discuss new applications for their equipment (with appropriate confidentiality agreements in place) and their interest in sharing some development costs or forming strategic partnerships.

For new processes and applications, it is usually necessary to perform testing and trials involving the supplier, especially if they are being asked to provide customization and/or guarantees. It is best to run trials with multiple potential suppliers to maintain a competitive position ahead of negotiating a development agreement that provides clarity through the commercialization of the technology.

It is important prior to these trials to have clarity regarding any intellectual property that may be generated. When negotiating these agreements, keep in mind that many suppliers are themselves technology developers who need to protect their own intellectual property and freedom to operate. You will likely need to be more flexible and accommodating to work with them, but this may be worthwhile to leverage their expertise.

11.2.3.7   New vs. Used vs. Refurbished Equipment

When first embarking on a piloting project, your company will almost certainly be dealing with new orders of magnitude of capital expenditure. Compared to bench-scale equipment, chemical processing equipment can be expensive, and you may look to alternatives to save money. One such avenue is used and refurbished equipment.

Buying used processing equipment from a wholesale supplier is typically not so similar to buying a second-hand car from a dealership as to buying a car from a scrap yard. Used equipment is usually sold "as is" with no guarantees on performance or even baseline functionality, but often for 10–20% of its original purchased cost. The equipment may also have been disassembled for shipping to the wholesaler and stored in a warehouse, under tarps, or open to the elements for years.

With this in mind, when purchasing used equipment from these suppliers you should have experienced mechanical and electrical personnel or contractors available with good working knowledge of the item in question. You should also budget substantial time and money to get the equipment to a working condition. For example, if a large dryer was previously controlled through a plantwide control system, it will need to be set up with a completely new control system for the pilot.

When budgeting, do not forget to include cost for load-out and shipping, which can be a substantial fraction of the total capital cost. This is especially true if the equipment is purchased from a liquidation auction of a decommissioned plant. In these cases, safe removal of a few pieces of equipment can cost tens of thousands of dollars in rigging, loading, and packaging.

Refurbished or reconditioned equipment is a significant step up in terms of baseline functionality but also in terms of price. Reconditioned equipment may sell for ~50–75% of its original value, will typically be backed by a limited warranty from the vendor, and can be counted on to function on delivery. However, it is likely that only limited models will be available, so you may have to make compromises on the operational side to accommodate.

The availability of used and refurbished equipment often changes quickly, sometimes by the day if it is a piece in high demand. This makes it risky to build a pilot project around a specific piece of equipment until it is secured.

It is also important to remember that actual installation costs for used equipment will be similar, sometimes higher, than for a new piece of equipment, so the installation factors discussed in Part 1 cannot be used. Rather than cost, the main advantage of used equipment is that it may be available quickly, and in some cases may be able to improve the overall pilot plant development timeline by several months.

11.2.3.8 Materials of Construction

Materials of construction can be an important driver in the economics during scale-up as metallurgy substantially impacts the cost and availability of equipment and instrumentation. Often only specialized vendors will offer exotic and expensive alloy options, with lower competition leading to even higher prices.

For more common fluids and applications there may be substantial data available for alloy selection. However, caution is required for new processes as corrosion rates may be dramatically impacted by changes in impurity profile. In the absence of available data or experience, similar fluids can be considered for initial guidance, but experimental testing of different alloys across a range of concentrations and temperatures expected is a wise course of action. This may include longer-term corrosion coupon studies.

The selection of materials for a pilot plant should consider cost and robustness as well as the importance of gaining valuable data for the commercial plant. Sometimes the optimal solution is a mix of different materials. Nonmetal options should also be considered particularly for more corrosive conditions, as they are often lower cost. However, they are typically limited to much lower temperatures and pressures than metals and are typically more easily damaged by use or impact.

It is also important to recognize that materials in different services can tolerate different amounts of degradation. For instance, a thick-walled tank or pressure vessel may be able to tolerate a corrosion rate of 0.5 mm yr^{-1} over a 10-year lifetime by using a 6 mm corrosion allowance (extra thickness to compensate for corrosion losses). This is clearly not possible for a heat exchanger with tubes less than 2 mm thick, where materials are often selected to achieve corrosion rates of 0.1 mm yr^{-1} or less. Likewise, for nonmetals some swelling may be acceptable in one application such as a static gasket, but not in a more dynamic application.

It is also worth noting that some processes may be sensitive to low concentrations of dissolved metals or other corrosion products. Thus, materials deemed acceptable from an equipment lifetime point of view may actually be inappropriate. Further, some corrosion mechanisms generate hazardous and flammable materials that can accumulate to dangerous or explosive concentrations in tank headspaces, piping high-points, and other poorly ventilated locations. As an example, hydrogen is often formed when many common metals are corroded by acids. It is also not only the materials in direct contact with the process that need to be considered. Other materials such as supports, paint, insulation/cladding, concrete/coating, etc., should be suitable for periodic contact due to leaks, spills, or drainage.

Extra care is required for processes producing food and food ingredients. Depending on the exact service, this may require special surface finishes and/or material certification that it is noncontaminating to the process through to direct food contact. Some equipment/materials may also need to withstand periodic sterilization or clean-in-place (CIP) using suitable chemicals.

Consultants with relevant expertise may be worth their weight in gold (or tantalum!) if they are able to suggest suitable, lower-cost materials or help avoid a costly mistake in selection that would otherwise require subsequent rework with associated costs and delays.

11.2.4 Pilot Plant Operation and Location

11.2.4.1 Avoid Unnecessary Complexity

Technology development is hard enough, so try to keep things simple whenever possible. Maintaining a simple pilot reduces capital and operating costs and leaves you well-positioned to achieve your goals in the face of unforeseen issues. It is not usually necessary to build a smaller-scale version of the entire commercial process. In practice, this poses substantial operating challenges. Focus on piloting the units that are most important.

Simplification can also be attained by avoiding unnecessary interconnecting parts of the process. Tying systems together can make the pilot less reliable, flexible, and reconfigurable. For instance, it is typically not necessary or cost-effective to incorporate heat and pressure recovery in pilot operations. These are generally fairly well understood and simple to account for in mass and energy or cost models such that they can often be omitted without excessive risk to the commercial plant. Where fouling or extreme temperatures are considered important to study, the pilot can still be designed to investigate these without actually linking one part of the plant to another.

Focusing on steady-state operations can also dramatically reduce pilot complexity. Initial start-up and shutdown operations create transients that are very different from the process that has settled into its steady-state operation. For example, off-spec products and intermediates produced during this transient time tend to dominate. As these will disappear during steady-state commercial operations, piloting these phases of operation is often not necessary.

Think about recycle streams as well for areas of simplification. Understanding/evaluating recycles is one of the primary goals of piloting. However, it is not always necessary to incorporate all recycles, especially during early stages of development. In some cases, instead of implementing a full recycle stream, an intermediate might just be vented and fresh chemical used as feed (in place of the recycle stream). However, it is critical to ensure that if there are important impurities present in these recycle streams, they are added (spiked) into the process at a later stage.

These are just a few of the potential simplifications and cost savings measures that might be employed by an engineer with experience developing new processes. These will act to minimize risk to the commercial plant design while ideally remaining within budget.

11.2.4.2 Pilot Location/Integration

The cost, complexity, and flexibility of a pilot can be impacted significantly by its location. There are a number of different options in terms of where the system is set up and how it is integrated into existing infrastructure. These include using a dedicated facility, a third-party site, a mobile unit, or a unit integrated with an existing industrial unit. Each of these have pros and cons as described in the following table:

Pilot Install Concept	Pros	Cons
Dedicated facility – a new facility set up for the purposes of the pilot.	• Built for purpose, so features can be tailored to the needs of the project. • Trade secrets and know-how can be easily maintained. • Own staff can be used for operations, reducing operating costs. • Flexibility in hours of operation.	• Expensive – all utilities must be generated for the pilot. • Time is needed for facility setup. • Less access to industrial experience.
Third party site – an existing research and development center (private, university, or government run).	• Ties into existing utilities. • Experienced staff available to assist in troubleshooting. • Simpler insurance and permitting (through host).	• Space rental and labor charges add to ongoing operating costs. • Confidentiality agreements must be put in place to protect IP. • Compliance with host work and safety procedures may add to cost and timeline.

Pilot Install Concept	Pros	Cons
Mobile unit – a system designed to be shipped to different prospective client sites. Typically built as a stand-alone system with a minimum of integration work at site.	• The same system can be trialed for multiple prospective clients. • Minimum ongoing space rental costs. • Ideal for small scale pilot systems.	• Shippable systems have tight build envelopes, and so require more up-front engineering and are harder to reconfigure on the fly. • Physical limitations may define the achievable capacity of the pilot.
Integrated with an industrial plant – a process intended to work as part of an existing industrial process. It may need to be integrated into a slip-stream of an industrial plant in order to conduct a representative trial.	• Tie into existing utilities. • Industrial expertise available. • Opportunity to show technology directly to client in its own facilities.	• Engineering work to integrate with a plant is a major expense. • Operational flexibility will be limited as procedures will be carefully vetted by plant safety personnel. • Operators may not have experience with development projects. • Any shortcomings in the design will be directly apparent to the client – even minor issues that are easy to fix can hurt client confidence in the technology.

11.2.4.3 Pilot Unit Operating Philosophy

A major driver that will affect the design is the way in which the plant is intended to be operated. At one end of the spectrum is operating the pilot unit completely in a manual or "hands-on" manner. This involves a unit that is continuously manned and operated with manual transfers and control systems. For example, such a system could consist of discrete pieces of equipment, operated with manual valves, push buttons, visual gauges, dials etc.

The main advantage of this mode is that it is relatively inexpensive and does not rely on complex electronic control systems, which can take time to set up safely and troubleshoot. However, more labor is needed to operate, which increases your operating cost, and it can be hard work for operators, leading to retention challenges. There is also greater risk of contact between personnel and process chemicals, potentially making it unfeasible for more hazardous processes.

At the other end of the spectrum is a completely automated system. This involves a considerably more complex system that operates with microcontrollers or programmable logic controllers (PLCs). The advantages are that no human intervention is required, and the system can be run with much more precise control. Of course, such a system is much more difficult to design, install, and adapt, leading to higher capital cost and may be overkill for relatively simple processes.

This selection will depend on the objectives of the pilot, its capacity, and the manner in which it is integrated. Generally speaking, the decision will be based on satisfying the trade-offs between capital and operating costs as well as between automation and flexibility. There is of course a continuum of pilot setups across these extremes, and many hybrid systems can be employed to meet testing requirements.

11.2.5   Conclusion

The intent of this chapter was to give you a broad overview of how to advance your sustainable chemical technology from the bench to the pilot scale on the way to a commercial reality. This brings a whole new set of challenges that require specialized engineering expertise to successfully navigate.

Editor's Note

Don't worry if you didn't grasp all of the technical details or engineering concepts presented here. Unless you have studied chemical engineering or something similar, there isn't any reason for you to have been exposed to these concepts before, and it is a different way of thinking. However, it is critical that you be aware of the concepts at a high level as it will allow you to avoid missing important steps and will help you communicate effectively with your engineering team. It will also allow you to evaluate and select your engineering partners and build a strong and diverse team. This is critically important at the scale-up and piloting stages and ultimately to the success of your venture.

12

Raising Investment/Financing

Matthew L. Cohen

Pangaea Ventures, Phoenix, AZ, USA

12.1   Introduction

Adena Friedman, the president and CEO of NASDAQ, told the graduates of Vanderbilt University's Owen Graduate School of Management in a May 2017 commencement address that "Ideas are only as good as your ability to communicate them" (https://www.nasdaq.com/articles/adena-friedman-gives-commencement-address-vanderbilt-owen-2017-05-15). The same can be said for forming a company to commercialize a sustainable chemistry innovation. No matter how clever your approach is or how large of a market you could disrupt or how much of a positive environmental impact you could have, if you cannot effectively communicate your idea, you won't attract the resources required to realize your vision. To obtain these resources, you're going to need to know who to talk to, how to talk to them, and when to initiate these discussions. Fundraising is an essential, but often overlooked, component of the commercialization journey and should be considered from the earliest stages of bringing an innovation out of the laboratory and into a start-up with commercial aspirations.

As you read this chapter, there are many hundreds of sustainability start-ups attempting to commercialize disruptive technologies and many more thousands of promising ideas that could change the world. If there is one overarching point to consider when contemplating fundraising for a start-up, it is this: investors don't like to simply fund good ideas nor do they look to invest in science projects. Of course, there are exceptions, but if you're reading this, I'm willing to bet that you aren't an exception.

Famous people and people who have made boatloads of money for their investors in past start-ups can maybe get funded on the merits of their ideas alone. For the rest of us, I'm afraid a brilliant idea is not equivalent to a fundable opportunity. David Rose, the CEO of Gust and Chair of the New York Angels, sums it up well. "While a good idea is usually a necessary ingredient for the formation of a good company, it is not sufficient by itself for any serious investor to fund. Why? Because there are also other good ideas out there, some of which have already been developed, tested and put into practice, thus decreasing the amount of risk an investor will be taking. The bottom line is that ideas by themselves are simply not fundable by professional investors" (http://blog.gust.com/how-do-i-get-in-touch-with-investorsfunds-with-just-an-idea-and-no-product).

How to Commercialize Chemical Technologies for a Sustainable Future, First Edition.
Edited by Timothy J. Clark and Andrew S. Pasternak.
© 2021 John Wiley & Sons Ltd. Published 2021 by John Wiley & Sons Ltd.

Raising money for sustainable chemistry-based innovations presents unique challenges. Unlike in software, where a few people can hypothetically crank out a minimum viable product (MVP) in a garage on a weekend, sustainable chemistry innovation requires capital equipment, lab facility infrastructure, and materials, and these aren't cheap. This chapter will delve into these challenges as well as some of the investment drivers that can counteract the often-formidable obstacles. There certainly is no one-size-fits-all solution to how a company should (or shouldn't) raise investment. Understanding the financing landscape is helpful if you plan to try to navigate the murky waters of commercialization and investors' decision-making processes.

Beyond the discussion of the investment vehicles themselves, many of the concepts related to financing that you will encounter in this chapter have been previously discussed (sometimes in greater depth) in previous chapters. However, this chapter takes the perspective of a financier in the venture capital industry who has evaluated and invested in a wide range of early-stage chemistry-based companies. It is more high-level, with an emphasis on understanding and evaluating the investment landscape.

12.2 Main Investment Sources

The financial resources necessary to commercialize sustainable chemical technologies come in different forms. Whether it's a nondilutive grant to get things going, debt to fund the major capital expenditures for a pilot-scale plant, or something in between, each type of funding has its positives and negatives. The relative mix of each will almost always evolve over the lifetime of a start-up. These different kinds of investment sources will be defined, compared, and contrasted. Note that additional information about grants and strategic partnerships can be found in Chapters 5 and 9, respectively.

12.2.1 Grants

At the earliest stages of commercialization, grants are the most common (and often the only) form of monetary support available to entrepreneurs. I will be using the word "grant" as a catchall term for related nonequity, nonpartnership funding such as business plan contests and pitch competitions. The money received from grants does not need to be paid back and, unlike direct equity investment, does not convert to any ownership in the company that received the grant funding. Grants are a key enabler for sustainable chemistry start-ups, and it is rare to progress to sizable equity funding rounds from venture capitalists (VCs) without them.

12.2.2 Strategic Partnerships

Large, established companies look to engage in "win-win" engagements with start-ups as each party can often bring something unique and valuable to the table. The large company can provide capital or market access, and the start-up can usually contribute innovation and nimbleness to bring something new to market. Partnerships come in many forms and flavors, and the attractiveness of the arrangement must be judged on a case-by-case basis. Two common forms of strategic partnership funding from a large industrial firm to a start-up are nonrecurring engineering (NRE) and joint development agreement (JDA) payments (https://physicsworld.com/a/four-ways-to-fund-a-start-up); both are important slices of the funding pie for chemical technologies.

12.2.3 Equity Investment

Angel investors, VCs, corporate venture arms, family offices, and private equity firms typically support companies via equity investment. In exchange for contributing cash to the start-up, investors receive shares in the company and thus own a portion of it. In its simplest form, shares are issued at a set purchase price ("priced equity round"), and ownership percentage is determined by the total number of outstanding shares. However, it is quite common for earlier stage start-ups to raise capital via convertible notes or Simple Agreements for Future Equity (SAFEs), which are a bit different. As opposed to granting shares immediately to the investor, both convertible notes and SAFEs are securities where money invested in a company converts into shares at a future date. Conversion typically happens at a financing milestone like a Seed or Series A funding round. There are lots of great resources that go into detail on this topic (Alexander Jarvis' site is one and is a handy resource for entrepreneurs and investors alike: https://www.alexanderjarvis.com/differences-between-safe-and-convertible-notes).

Angel investors are most commonly high-net-worth individuals investing out of their own pocket. This contrasts with VCs who invest out of funds raised with external investors called "limited partners" (LPs). There are also family offices where a wealthy family's money is managed by investment professionals. In many ways, family offices are like VCs, but instead of many external LPs that commit capital to a fund, there is just one LP – the "family." Family offices are becoming a more important component in the equity funding mix for sustainable chemistry start-ups because many of these family offices have mandates to support companies making a positive impact on the world.

It is also worth touching on another type of VC known as "corporate venture capital" (or CVC, sometimes also referred to as "corporate venturing"). In the chemical technology sector, CVC groups are involved more frequently in deals compared to other heavily funded sectors like software. Corporate venturing is defined as the practice of large, established companies directly investing into external start-ups (https://corporatefinanceinstitute.com/resources/knowledge/finance/corporate-venturing-corporate-venture-capital). For example, both BASF and BP operate CVCs. These multibillion-dollar companies are often strategically driven to invest in innovative start-ups, and they are not just seeking simple financial returns. They may be looking to expand their own product lines or to find ways to increase sales of materials required by the start-up or its customers. Some CVCs are structured to invest directly from the corporate parent's balance sheet, while others have external funds with the corporate parent as the sole LP. There are also some that operate as an external fund with multiple third-party corporate LPs. For a primer on the CVC landscape and how to engage them, you can refer to Scott Orn's and Bill Grownley's TechCrunch article (https://techcrunch.com/2020/05/26/how-to-approach-and-work-with-the-3-types-of-corporate-vcs). CVC investors are commonly referred to as "strategic investors," as they are investing for additional longer-term positional gains beyond financial returns.

Finally, for more established start-ups, another constituent of the equity funding mix is private equity (PE) firms. PE firms tend to write larger checks compared to VCs and to invest in later-stage, less-risky companies that already have substantive market penetration and are turning a profit. Typically, PE firms target larger ownership percentages compared to VCs and often leverage the equity capital they contribute to a company directly with additional debt in order to take controlling ownership stakes and/or to fund roll-ups (acquiring and merging several smaller companies in the same market).

Regardless of the flavor of equity financier, their ultimate goal is to invest in a company that can return a lot more money than the initial invested capital. As such, equity investors typically target "exits" for their portfolio companies. Exits can come in many shapes and sizes, but the main two are mergers and acquisitions (M&A) and initial public offering (IPO). In an acquisition exit, one company purchases another and typically pays cash and/or stock to do so. The shareholders of the acquiree ideally get paid substantially more than they previously invested. This is the most common form of exit for sustainable chemical start-ups. Mergers, where two companies are combined as relative equals, also happens but is much less common. In a similar vein, IPOs are also rare for start-ups, but a select few are able to go public and offer shares in a new stock issuance to the public on an exchange like the London Stock Exchange or NASDAQ. Once publicly listed, equity investors as well as the founders and shareholding team members can sell their shares to generate a return.

12.2.4 Debt

Debt, typically in the form of a loan or series of loans, is another funding source, and one that can be complementary to equity investment. Unlike dilutive equity financing, debt is not by definition an instrument of dilution (though in practice, debt terms sometimes include warrant coverage, which is dilutive). The flip side of this is that debt must be paid back. Regardless of whether things work out, a company that takes on debt is on the hook to pay back that money.

Within the debt world, there are two broad classifications – traditional debt financing and venture debt. Traditional debt financing typically is only for lower-risk, later-stage companies that can demonstrate at least a few years of profitability. Institutional lenders will sometimes use the word "bankable" to describe these kinds of companies. Start-ups typically don't get access to much traditional debt financing as most aren't bankable given their relatively nascent stage. Venture debt, on the other hand, is for companies that lack the assets or cash flow for traditional debt funding but are seeking additional capital to grow. It is common for venture debt to be added in conjunction with or soon after an equity investment round. "When utilized appropriately, venture debt can reduce dilution, extend a company's runway, or accelerate its growth with limited cost to the business" (https://www.kauffmanfellows.org/journal_posts/venture-debt-a-capital-idea-for-startups).

Equipment financing or equipment leasing is a related debt concept and worth mentioning here due to its prevalence and usefulness for capital-intensive chemical technology start-ups. There are two common use cases. In the first, a lessor like a bank or a specialty finance company puts up capital to purchase a piece of equipment for the lessee, and this capital is secured by the equipment itself. This means a start-up can add a relatively expensive reactor, characterization tool, or other piece of equipment without paying cash up front to obtain it. The second scenario is called an equipment sale-leaseback, in which a company that already owns the equipment generates working capital by leasing it back to the lessee with the equipment itself as collateral on the loan. If the company defaults on the loan, they may lose access to the equipment since the equipment is technically now owned by the leasing company. Of course, this isn't a good outcome, but at least it's limited to the equipment itself. Defaulting on venture debt differs in a nontrivial way; if you're unable to pay back venture debt, the lessor may try to recoup its investment by forcibly selling off other company assets including its intellectual property (IP).

When a start-up has reached a stage of predictably generating monthly recurring revenue (MRR), debt can also come in the form of revenue-based financing (RBF loan) and receivables financing (aka "accounts receivable [A/R] factoring"). Revenue-based financing provides capital for growth in return for a percentage of monthly revenues. Unlike traditional term loans, RBF loans have more flexible repayment schedules as they're tied to the ups and downs of a company's monthly net revenue. A/R factoring is a debt instrument where a business sells – at a discount – its accounts receivable (e.g. invoices not yet paid by their customers) to a lender in order to get an immediate cash payment. This can be helpful for alleviating working capital issues.

To complete this overview on debt, it should be noted that there are some company-friendly programs and debt facilities that are quite applicable to sustainable chemistry start-ups. Economic development programs focused on regional job creation often have more favorable terms compared to traditional debt sources. There are also debt programs specifically tailored for companies with positive environmental, social, and governance (ESG) profiles. The Business Development Bank of Canada's (BDC) Cleantech Practice is one example (https://www.bdc.ca/en/about/what-we-do/cleantech-practice/pages/default.aspx). In the challenging environment caused by the COVID-19 global pandemic, BDC Capital also created a special program for Canadian start-ups to match, via a convertible note, financing raised through qualified existing and/or new investors (https://www.bdc.ca/en/bdc-capital/venture-capital/strategic-approach/pages/bridge-financing-program.aspx). Governments around the world and the public alike are pushing their ideological desire to support opportunities with strong environmental impacts, and we are starting to see the numbers reflect this desire. BloombergNEF reported that "global green and sustainable debt volumes hit close to $500B at the close of 2019, more than double the figure of two years previously" (https://www.privatefundscfo.com/esg-financing-cheaper-source-of-debt-or-too-expensive-to-ignore).

12.2.5 Bootstrapping (Using Your Own Money)

All of the aforementioned categories rely on money or resources from outside of the company for commercialization and growth. Bootstrapping represents an alternative approach where an entrepreneur creates and grows a company from personal finances and/or operating revenues. If you can grow your business organically by reinvesting the revenue from the sale of your product or service, why bother with external investors at all? For sustainable chemistry start-ups, at least at their inception, it's pretty uncommon to be in the position to completely bootstrap your business. In a subsequent section, we'll circle back to the trade-offs between bootstrapping – heightened personal financial risk and slower growth – and raising various forms of outside investment.

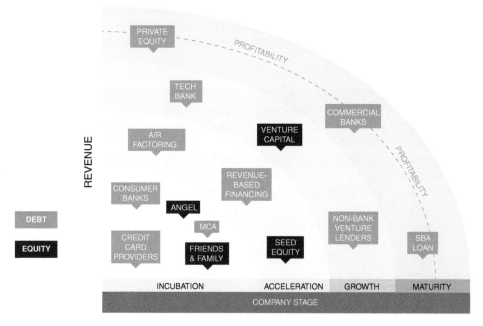

Figure 12.1 Available forms of financing. Source: From Lighter capital the startup finance playbook: Finding the right financing for your capital needs. https://www.lightercapital.com/the-start-up-finance-playbook/

12.2.6 Summary

Figure 12.1, adapted from a useful start-up financing options primer written by Lighter Capital called The Startup Finance Playbook (https://www.lightercapital.com/the-start-up-finance-playbook), depicts how financing sources depend on the company's stage of development. It isn't an exhaustive list complete with all the options discussed in this chapter nor is it specific to sustainable chemical technology ventures, but it is a helpful tool to visualize the myriad financing types and where each capital source fits with the maturation and growth of a company.

Now that the landscape has been elucidated and the requisite jargon defined, let's move on to considering the unique opportunities and challenges that investing in sustainable chemical technologies presents.

12.3 Unique Considerations for Investing in Sustainable Chemistry

As you might imagine, a start-up commercializing a sustainable chemical technology is quite a bit different from one developing a mobile app. Because of this, investors view playing in this area differently from other heavily funded areas like software or biotech. Investors typically juggle a number of different factors in making investment decisions including but not limited to perceived overall risk, types of risk (i.e. market, regulatory, scale-up, execution, etc.), and potential upside (exit type, size, and timing).

The following delves into specific investment drivers and impediments for investing in sustainable chemical technologies as well as how investing in sustainable chemistry stacks up against other fields where investors can deploy capital.

Sustainable chemical technologies: investment drivers and impediments

Investment Drivers	Investment Impediments
Responsible investing commitments	The (lack of a) "Green Premium"
United Nations Sustainable Development Goals	Commodity markets where cost is king
Consumers pushing brands to reflect their values	Reliance on policy or regulations
Circular economy and waste regulations	In-place infrastructure
The chemical industry's CO_2 problem	Historical subpar returns

12.3.1   Investment Drivers

Fortunately, there are some macroscale trends that act as investment drivers in sustainable chemical technologies. From individual consumers to companies both large and small, and even at national and international stages, there are distinct tailwinds pushing forward sustainability.

12.3.1.1   Responsible Investing

Considering the ESG factors in decision-making isn't a particularly new idea. However, the investment community has recently embraced it. The UN-supported "Principles for Responsible Investment" (PRI) is an international network of investors that incorporate ESG factors in their investment decisions (https://www.unpri.org/about-the-pri). With more than 2000 fund signatories that collectively have more than $80 trillion (yes, trillion with a "t") assets under management, responsible investing is not a passing fad or curiosity. Of course, only a small fraction of this gargantuan amount is specifically relevant for nascent sustainable chemical technologies. The good news is that there is a growing consensus among the investment community that investing both responsibly and to maximize long-term profits are not mutually exclusive. According to a 2020 Bain Global Private Equity Report, "What's changed in recent years is that vague discomfort about environmental and social issues has morphed into genuine alarm among investors and consumers. More and more people in both camps view existential global challenges such as climate change, plastics pollution, loss of biodiversity, deforestation, social inequality and water shortages as tangible threats. Conventional wisdom has it that these are 'millennial issues,' but the evidence suggests the concern is much broader" (https://www.bain.com/globalassets/noindex/2020/bain_report_private_equity_report_2020.pdf).

12.3.1.2   United Nations Sustainable Development Goals (SDGs)

In 2015, the UN General Assembly created a set of 17 global goals that were designed as a "blueprint to achieve a better and more sustainable future for all" (https://www.un.org/

sustainabledevelopment/sustainable-development-goals). Further details and examples are provided in Chapter 3. Many impact investment firms and funds have leveraged this framework to try to quantify the positive sustainability impact their portfolio is having on the world. One example of this is Pangaea Ventures' Annual Impact Report (https://www.pangaeaventures.com/impact). For Pangaea, evaluating investment opportunities with an impact-focused lens is one way to size up disparate opportunities since our fund can support only so many companies. Tracking portfolio company progress against some of the UN SDGs is a helpful barometer of progress and one that isn't often captured in traditional milestone reporting formats. Considering the potential SDG impacts of a new investment opportunity helps guide decision-making and (ideally) fund performance. Entrepreneurs who can convincingly articulate why their technology will win and why that matters in the context of the UN SDGs should have an easier time raising money from all sorts of investors.

12.3.1.3 Consumers Pushing Brands to Reflect Their Values

Consumers have more choices than ever for the products they buy. Increasingly, brands are feeling the pressure to make products that reflect the values of consumers to differentiate their product offerings and their brand images. That means that producers are devoting a lot of resources to remove hazardous materials or adopt safer chemistries and to reduce the carbon footprint of their products and company operations. One example is the lengthening of manufacturing restricted substances lists (MRSL), which put pressure on brands and their supply chain partners to find safer, sustainable alternatives. Some of the world's largest apparel companies are turning to sustainability to separate them from the pack. Nike's Move to Zero effort is a great example (https://www.nike.com/sustainability). This trend serves as a nice tailwind to implement novel sustainable chemical technologies, at least in consumer-facing sectors such as clothing, apparel, home care, and food packaging.

12.3.1.4 Circular Economy and Waste Regulations

There is growing consensus that the chemical industry's traditional linear economic model of "take, make, use, and waste" is not sustainable (https://www.strategyand.pwc.com/m1/en/ideation-center/research/2019/circular-economy/circular-economy.pdf). While firms can often gain a competitive advantage by embracing the tenets of a circular economic model, regulators are starting to enact policies to forcibly push the industry in a more sustainable direction by encouraging circular economy practices where waste is repurposed as a feedstock. We're already starting to see some results: for the first time in Europe, there was more collection and recycling of post-consumer plastic waste than the plastic that went to landfill in 2016 (https://www.plasticseurope.org/en/newsroom/press-releases/archive-press-releases-2018/european-plastic-waste-recycling-overtakes-landfill-first-time). In addition, in 2018 the EU updated its 2015 circular economy action plan to increase use of recycled plastic by mandating recycled content minimums and waste reduction measures (https://ec.europa.eu/environment/circular-economy/pdf/new circular_economy_action_plan.pdf).

On top of this, China, once by far the world's largest importer of recycled waste, changed course with its 2018 "National Sword" policy. This resulted in plastics imports in China dropping by 99% (https://e360.yale.edu/features/piling-up-how-chinas-ban-on-importing-waste-has-stalled-global-recycling). This has caused major shocks to the supply chain in places like the United States, Canada, and Australia. Companies in these countries have been scrambling to find alternative solutions to their waste problems. Combined, these policies present a great opportunity for novel chemical technologies to solve challenging compliance issues for the chemical industry.

12.3.1.5 The Chemical Industry Has a CO_2 Problem

The Organisation for Economic Co-operation and Development (OECD) predicts that the amount of materials consumed globally is projected to more than double from 79 Gt in 2011 to 167 Gt by 2060, and waste is projected to increase 70% by 2050 (https://informaconnect .com/circular-economy-chemical-regulations). Tying in to this growing demand for chemicals and materials, the International Energy Agency (IEA) reports that the chemical sector is the largest global consumer of both oil and gas and is the third largest emitter of carbon dioxide (https://www.iea.org/reports/tracking-industry/chemicals) after steel and cement. Taken together, these three industries represent a gigantic opportunity for sustainable chemistry. The chemical industry doubly impacts climate change as one must consider both direct emissions from operations as well as the environmental impacts of the manufactured intermediates and products.

While the challenges are immense, there are already some hopeful signs of change. Novel chemical technologies that can improve efficiencies and lower costs should help continue these encouraging trends and are necessary if the chemical industry is to comply with the IEA-based "sustainable scenario" for 2050. Specifically, this scenario means that the chemical industry will need to "quadruple its size to more than $18.7 trillion" by 2050, "while halving its total CO_2 emissions" [1]. Breakthrough technology innovations will be absolutely critical to accomplish this ambitious goal!

12.3.2 Investment Impediments

I think most investors would agree with the notion of making the most money while doing as much good as possible. Unfortunately, it's rarely that simple and having our (Earth Day–themed) cake and eating it too is more aspirational than realistic. Many factors present significant hurdles to overcome to successfully finance the development of chemical technologies for a sustainable future. These are covered in the following sections.

12.3.2.1 The "Green Premium"

Everybody cares about the environment... until it negatively impacts their wallet. McKinsey & Company conducted a survey in 2012 on whether consumers would be willing to pay a premium for "green" aka sustainable products (https://www.mckinsey.com/business-functions/sustainability/our-insights/how-much-will-consumers-pay-to-go-green). The good news is that 70% of consumer respondents across the US and EU said they'd be willing to pay an additional 5% for a greener product assuming it had the same performance of the less sustainable alternative. The bad news is that this percentage drops off precipitously as premiums increase, with willingness to pay going below 10% at a 25% price premium.

These numbers gloss over the inherent variability between industries (i.e. consumers are apparently willing to pay a higher green premium for packaging compared to electronics or furniture) and respondents (i.e. millennials may value "green" more than baby boomers), but the survey confirms the almost negligible premium most people are willing to pay for greener products. This finding is exacerbated by a related study McKinsey performed with company executives that concluded companies will only pay a premium for greener products and services in their supply chain if they can charge customers more down the line. The unfortunate conclusion is that it's generally a losing bet if you're thinking of starting a company that relies on a green premium to be economically viable.

12.3.2.2 Commoditization – Competing on Cost

Sustainability in the chemical industry can come in a wide array of forms. From leveraging renewable power generation and storage to drive energy-intensive chemical processes such as distillation to using novel biopolymers to target the replacement of petrochemically derived plastics, one commonality is that we're often dealing with commodities. And when we work with commodities such as electrons flowing in a circuit (i.e. electricity) or fungible monomers to make ubiquitous plastic containers, cost is the key differentiating metric.

To commercialize a novel chemical technology, the fundamental economics have to work. More than that, the economics often have to be favorable from the outset without leveraging massive economies of scale and efficiencies gained over decades of operating experience. Getting funding for a pilot plant that can't operate profitably at full capacity is a tougher sell compared to one that can break even at current commodity prices. The prevailing low (and in early 2020, negative!) price of oil has been the death knell for a number of higher-profile sustainable plastics companies in recent years including Bio-On (https://www.biofuelsdigest.com/bdigest/2020/01/07/bioplastics-maker-declares-bankruptcy), TerraVia (né Solazyme) (https://cen.acs.org/articles/95/web/2017/08/Algae-products-specialist-TerraVia-goes-bankrupt.html), and BioAmber (https://cen.acs.org/business/biobased-chemicals/Succinic-acid-maker-BioAmber-bankrupt/96/i20). Competing strictly on cost is a race to the bottom, but it's often the operating reality in chemical technology commercialization.

12.3.2.3 Reliance on Policy/Regulation

Most investors do not like investing in companies where the stroke of a pen following a policy or regulation decision can make or break a business. Investors would much rather take on risks they can influence such as execution risk compared to policy risk. If the executive team fails to do what they said they were going to do, the team can be augmented or changed. However, if a new law or executive order is passed that has a severe negative impact on the prospects of a business, there's typically not much recourse. Because of this, policy risk is an important consideration for investors in this space. In the US, we've seen several environmental policy swings between executive administrations. This uncertainty can make it difficult to fund long-term capital investments, especially for more conservative institutional investors with deep pockets. It's tougher to justify investment if there is a risk that the modeled five-year return on investment (ROI) on a project suddenly jumps to 15 years because the cost of carbon dropped due to a new administration's policy changes.

12.3.2.4 In-Place Infrastructure

Tesla and Panasonic's Gigafactory in Nevada is an audacious undertaking to reduce the cost of lithium-ion batteries and packs for electric vehicles (EVs). It could be the key enabler for EV production and operation to be lower cost compared to gasoline-powered internal combustion engines. However, the capital investment for this is similarly audacious, with a multibillion-dollar price tag. In 2016, Panasonic raised $3.86 billion in corporate bonds primarily to support this effort (https://www.reuters.com/article/us-panasonic-results-tesla-idUSKCN1090WZ). Therefore, if you're commercializing a new better/cheaper/faster EV battery technology, it will need to be compatible with the manufacturing methods employed in the current infrastructure of the Gigafactory. The same argument can be made for greener processes in the petrochemical industry. To recoup their sizeable investments, companies are wedded to the infrastructure they currently have, so convincing them to make even minor changes is tough.

Too often, entrepreneurs focus solely on why their technology is better than the incumbent technology without proper accounting of the upstream and downstream ramifications of actually implementing it. Considering your technology within the larger supply chain (as detailed in Chapter 8) as well as the value chain framework originally developed by Michael Porter more than 30 years ago (https://www.isc.hbs.edu/strategy/business-strategy/pages/the-value-chain.aspx) is especially important for sustainable chemistry companies. Inertia is an immutable force of nature. Overcoming the inertia of in-place, costly infrastructure is no small feat and requires a truly transformational innovation instead of an incremental improvement.

12.3.2.5 Historical Returns

The MIT Energy Initiative released a report comparing the performance of every medical, software, and cleantech company that received its first round of VC funding between 2006 and 2011 (http://energy.mit.edu/publication/venture-capital-cleantech). Unfortunately, the results are quite sobering; $25 billion of funding went into cleantech start-ups, and the VCs lost more than half their money! The investing boom times of the mid-2000s quickly went bust during the Great Recession. While other VC sectors were hit hard, they recovered fairly rapidly while cleantech investing did not. Most people concluded that cleantech's risk/return profile just doesn't pencil out for VCs to make large, risky bets. While there are encouraging signs of a cleantech VC resurgence (though most have now dropped the term "cleantech" due to the connotation), the historical poor performance of investments in sustainable chemistry innovations continues to loom menacingly in the recesses of the minds of would-be investors (https://www.forbes.com/sites/robday/2019/07/22/cleantech-venture-capital-is-back-early-numbers-say-yes/#2699fa357d00).

12.3.3 Comparison to More Heavily Funded Areas

It can be helpful to compare the commercialization ecosystem for sustainable chemistry with other well-funded industries like software or biotech. I have made some broad generalizations, and there will always be exceptions, but the following metrics offer a comparative framework.

12.3.3.1 Capital Intensity

Commercializing chemical technologies for a sustainable future is a lofty goal and challenging to do successfully. It's a poster child for the moniker "hard tech" aka "deep tech."

Hard-tech start-ups by definition are working on difficult technology problems that have game-changing potential. Unfortunately, the flip side of this is the capital needed for the development work to get to a commercial product is usually much higher than required by a software company, for example. Initial costs are often high just to prove out the idea. Worse yet, costs only get higher during scale-up to prove out the economics. Application development, testing products in real-world conditions, and building pilot units or plants are not cheap activities. There are dozens of skeletons of next-generation lithium-ion battery aspirants who ran out of fuel (money) lining the road that serve as cautionary tales.

Since most sustainable chemical technologies are destined to compete at least partially on cost in commodity-like markets, leveraging economies of scale to drive down costs is often a necessary evil to achieve success. Remember that economies of scale require achieving scale in the first place, and this in turn takes money. Compared to developing an MVP or scaling for a software product, the capital intensities are staggeringly different.

12.3.3.2 Commercialization Time Horizons

While the tenets of the lean start-up methodology (http://theleanstartup.com/principles) are great to keep in mind for any entrepreneur regardless of industry, sustainable chemistry start-ups often struggle because the iteration cycles can be quite slow. A year of market discovery, three more for materials/chemistry development and optimization, two more to build a pilot plant, and another two for full commercial production is not an unheard-of development timeline for a chemical technology start-up. Five or more years from inception to first product revenue is not the exception here – it's the norm. And while you toil away in the lab going after a moving target, your competitors will not be sitting idly.

The time value of money dictates that an investment that can scale and exit in less time is more valuable than an investment that achieves a similar scale and exit in a longer time frame. To make matters worse, a low growth rate may heavily reduce the terminal exit value. Exponential growth is much more common for a new mobile app on the Google Play store than it is for a novel chemical technology.

VC funds are usually set up to invest in companies, help them scale, and achieve exits within seven to 10 years. Unfortunately, for many chemical and materials technologies, the time required to go from lab discovery to exit can be longer than this. Thus, it can be challenging to make a traditional venture funding model fit. Consequently, many of the traditionally structured VC funds (aka "closed-ended" investment vehicles) have stopped investing in the space or repositioned to focus on later-stage companies closer to exit. Some VC funds are set up as "evergreen" or "open-ended" funds, so they're not as time-to-exit constrained as traditional closed-ended VC funds. Breakthrough Energy Ventures (www.b-t.energy) is one relevant example. Evergreen funds can help alleviate the fund structure incompatibility issue, but even for them, time is money, so how long it will take to scale is still a significant consideration.

12.3.3.3 Risks

Sustainable chemistry entrepreneurs face similar risks to entrepreneurs commercializing other types of technology innovations. These include, but are not limited to, the following:

- Market risk (do people truly want what you have?)

- Competitive risk (is what you have that much better than the incumbent solution to justify switching?)
- Technology/execution risk (can you actually make the technology work in practice?)
- Financial risk (will you run out of capital before the business becomes self-sufficient?)
- Regulatory risk (could a change in government policy derail your business model?)

To sell a novel medical device, there is a known regulatory pathway and reimbursement code. If all of the well-documented steps are done successfully, the FDA (or analogous organization) will approve the product. The company can then start selling it and be reimbursed at a predetermined rate. Unfortunately, it's often less straightforward for sustainable chemistry technologies. Relevant regulations and policies can differ between regions as well as national levels, as described in Chapters 6 and 7. Regulations covering air quality standards, CO_2, or VOCs can also change significantly from one administration to another.

12.3.3.4 Value Capture

Chemicals and materials are the underlying components of all physical goods. Unfortunately, firms that work in chemicals and materials sit far away from the end product in the value chain. Because of this, it can be challenging for a company commercializing chemical technologies to capture much of the value that they are enabling. The company that supplies the gases used to dope the silicon that is fashioned into the transistors that underpin the high-performance processor in a smartphone captures a vanishingly small fraction of the overall value of that smartphone. This is not a major issue faced by a biotech start-up with a blockbuster new therapeutic.

12.3.3.5 Exits

In biotech, it's pretty common for companies to submit an IPO in the midst of a Phase 2 trial or to command a $100 million+ acquisition offer before a new drug is even fully approved for use (and it may never be!). In IT, acquirers usually place significant premiums on growth and are willing to pay high revenue multiples.

In sustainable chemistry, you should expect none of this. Industrial sector acquirers are typically risk-averse and want to see more traction before making a large offer to buy a company. They'd rather pay a bit more in the future than take on risks now before an opportunity is fully operational. It is rare for a pre-revenue sustainable chemistry company to successfully submit an IPO. Therefore, the companies that do make it to an exit tend to be more mature than in other industries. Time to exit is longer, and exit sizes tend to be smaller. On the public markets, chemical companies are often valued at lower multiples compared to their IT brethren. This is the reality that sustainable chemistry entrepreneurs must face, and it can be a tough burden to overcome when trying to raise investment from generalist investors.

12.4 Financing Considerations

When evaluating the next stage in getting your company financed, consider the options available to you and the compromises that you're willing to accept.

12.4.1 Trade-Offs Between Investment Types

Grants get you going, but with strings attached.

At the onset of commercializing a novel technology, grants are a fantastic resource. They provide the super-high-risk capital needed to advance a clever idea into a prototype or validate a market hypothesis. They also don't require giving up a lot of your nascent company. This is nice because if you had to raise similar funds from a private investor at this stage, it would cost a large chunk of equity as valuations for very early/high-risk hard-tech start-ups tend to be pretty low. Furthermore, grantees typically retain full ownership of the IP developed during the grant program.

Grants can help build shareholder value because the less perceived risk on the table, the more value your company will be ascribed in a financing round. If a grant can support six months of necessary process development work, it will be easier to justify to an investor to pay a higher price for a piece of your business. Plus, getting a grant is a form of validation in the eyes of investors as the granting bodies typically have subject-matter experts reviewing the applications. Investors can and do lean on the technical due diligence performed by a credible granting body.

However, grants don't just grow on trees, and they tend to be competitive. There's a significant time cost to applying for grants, and time is precious for entrepreneurs. Of course, there's no guarantee of success either, though the odds are likely quite a bit higher versus pitching to a VC. There can also be a lot of time and effort needed for financial reporting after attaining the grant. This reporting process can be onerous for a small start-up. Many grants also restrict the ways in which you can use the funds. That means that costs for filing patents, marketing, or capital equipment may be ineligible.

The other thing to watch out for is grants can be defocusing. Many times, companies find themselves in situations where they can apply for a grant, but its purpose is to support a side project or a completely separate lower-ceiling opportunity with no tangible synergy to the company's main goal. A firm that has subsisted on various grants for 10+ years can be a red flag for an equity investor. Company culture can't be changed overnight, and the shift from an R&D, grant-focused organization to a product-focused, high-growth business can be painful. The stigma of being heavily reliant on continual small business grants can make it more difficult to attract angel or venture investment.

A final consideration on the topic of grants is matching funding. Many grant programs, especially the larger ones that can award $500 000 or more, stipulate that to draw down the funds from the grant, the company (or its investors) must put up matching funding. Sometimes it's a 1 : 1 ratio of grant funding to matching funding; other times the ratio is less depending on the rules of the particular grant. Matching represents both a challenge and an opportunity for start-ups. The challenge is obviously scraping together the resources to demonstrate the funding match. The opportunity is that a smaller amount of outside dilutive capital can be leveraged into more available resources for the company. Investors often view this positively as it gives a slug of free capital on top of their investment. Of course, this assumes that the grant program aligns with the start-up's mission. As such, it's pretty common for early-stage companies to initiate a fundraise after getting a grant in order to capitalize on the momentum and validation as well as to bring in the required matching funding.

12.4.1.1 Strategic Partnerships – The Devil is in the Details

Is it possible to craft a win-win partnership between a no-name start-up and a Fortune 500 company so that both parties can benefit equally and substantially? Sure it is, at least hypothetically. In reality, though, it can be quite tricky to structure a strategic partnership in a way that aligns incentives and meaningfully impacts all involved.

Similar to winning a grant, a strategic partnership with a brand-name company in your industry can bring some much-needed credibility. However, it can also scare away potential customers that are competitors of your partner. A start-up that has a strategic partnership in place with Adidas may find an unreceptive audience at Puma or Reebok. If the terms of a partnership make it near impossible to bring in capital from other investors, you may have signed away your company without even knowing it. Avoid getting subsumed into a subpar acquisition deal masquerading as a strategic partnership.

Strategic partnerships are often focused on solving the large company's problem. That means that large companies will push to have a tailored solution that meets their specific needs. This can lead a start-up down a path of spending a lot of time and effort solving problems that have limited applicability to the market as a whole. However, the knowledge a start-up acquires through this engagement can often lead to technical improvements that do indeed have broad applicability beyond the large company's specific use. The spillover benefits can be significant. Therefore, the way IP is treated in a partnership is a critical consideration. Here is as good a place as any to emphasize the importance of your company having good legal counsel.

12.4.1.2 VC Isn't a Panacea

Venture capital gets the lion's share of publicity when it comes to financially supporting high-potential start-ups. In reality, though, VCs invest in a surprisingly small percentage of nascent companies. Fundable data indicates that less than 1% of start-ups raise money from angel investors, and only 0.05% (i.e. 5 out of every 10 000 start-ups) are funded by VCs (https://www.entrepreneur.com/article/230011).

The cold reality is that the vast majority of companies do not fit the profile for a successful VC investment. Remember, VCs commonly are constrained by their investment fund structure, as funds have finite lifetimes. VCs strive to achieve big exits before the end date of their fund. Simply put, few companies are able to return to investors 10 times their initial investment in less than 10 years. That's a very high bar, especially for hard-tech start-ups where key milestones usually take longer than a software company and exit multiples are often lower compared to other industries. It's worth noting that even among companies that attract VC funding, the vast majority fall short of this 10x return goal. Regardless, you must be able to convince a venture capitalist that there's at least a theoretically viable pathway to achieve a big exit in a short time frame. This can be a tough argument to make, especially if you can't point to comparable high-profile exits of companies in your industry with similar business models.

Equity investment is a great way to add fuel to the fire, but it's not so great if you are using it to solve fundamental technology development issues. You can throw money at the problem, but that's not a guarantee that the problem will be solved, and it's definitely not a guarantee it will be solved in a time frame compatible with venture investment. In almost every case, start-ups don't raise all the money they will ever need all at once. Instead, money is raised in rounds, and there is an expectation that certain milestones will be achieved in order to get to the next round at a higher valuation (i.e. at more favorable terms for the entrepreneur or management team). If these key milestones are missed, the chance of being able to raise more money, at least on founder-friendly terms, decreases substantially.

For sustainable chemical start-ups, their development pace and ability to demonstrate value inflection points does not always align with the one or two years of runway that venture funding typically offers. To make matters worse, a company's net burn rate (the rate at which a company spends money in excess of income) normally increases after VC financing. This makes the company more reliant on further VC financing in the future until the company reaches profitability. As VCs want outsized exits, they characteristically push companies to grow as fast as possible. An aggressive growth plan often means higher spending and thus requires more time and scale to achieve cash-flow breakeven. Once you get on the VC rollercoaster, strap in and don't expect to get off mid-ride.

Venture investors want their portfolio companies to grow big and to grow fast. That means that they want their money to be used to attack a specific market opportunity. While your novel material or chemistry may be beneficial in a range of arenas, VCs will steer companies to be laser-focused on the area with the highest potential and probability for a large return. Combined with the higher burn rate after VC financing, it becomes more difficult for a venture-backed start-up to pivot and pursue a different avenue compared to a company that eschews venture dollars and is willing to grow more slowly with less financial backing. It also means classic "platform technologies" can be a challenge to get VC funded. Many shots on goal may mitigate risk for some companies, insulating them from events like a downturn

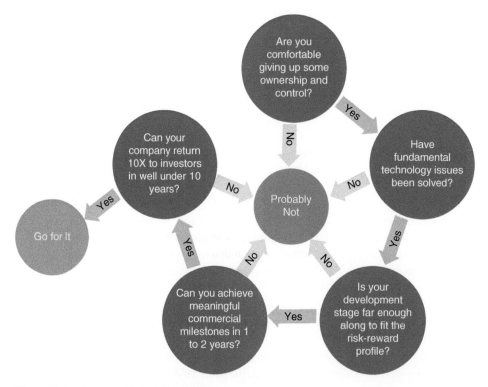

Figure 12.2 Company Fit for VC Funding.

in a certain market subsector. However, a start-up pitching to a VC that comes off as unfocused or still trying to nail down product-market fit is unlikely to attract much interest.

Figure 12.2 can be used as a quick guide to assess whether your company may be a good fit for VC funding.

12.5   Best Practices to Present Your Company to an Investor

Whether you pursue VC funding or other sources, there are some general practices that should be followed. The following sections offer suggestions on how to increase your chances of success.

Before a first meeting, do the following:

- Focus on the correct investors – do they invest in your market, in your geography, at your stage, etc.?
- Try to get a warm introduction whenever possible. Cold emailing, especially to venture capitalists who look at hundreds of start-ups per year, tends to be a fruitless endeavor. Fast Company's study showed a 1.7% response rate to cold emails (https://www.fastcompany.com/3036672/what-we-learned-from-sending-1000-cold-emails).
- Don't expect a busy investor to have spent more than 15 minutes looking at your company before an initial conversation.
- Don't overwhelm a potential investor with a 1000+ word email, a 30+ slide presentation, and/or a detailed business plan.
  - As an aside, a business plan is defined as a "written document describing the nature of the business, the sales and marketing strategy, the financial background, and containing a projected profit and loss statement" (https://www.entrepreneur.com/encyclopedia/business-plan). In practice, business plans as a stand-alone formal written document are being deemphasized in VC financing these days. While the content of a typical business plan is certainly relevant during the investment due diligence process, the information is more frequently split up across more narrowly focused documents rather than in one formally written business plan.
- Think about why this is the right time for an investor to support your company.
  - If others have tried and failed to disrupt the market you're pursuing, why will this time be different?
  - Are there external factors or tailwinds that may influence your chance of success?

During a first meeting, do the following:

It's cliché, but the goal of the first meeting is simply to get to a second meeting. This requires generating some excitement and demonstrating the potential mutual fit between your company and the funder's investment strategy.

- Focus on "why," not "how." Why is this opportunity going to the change the world and relatedly, why is the problem that you're addressing worth solving? If a financier is unclear on the "why," how you're doing it or going to do it doesn't really matter. For technical founders, this can be especially challenging.
- Be sure to have responses to questions like "Why are you and your team uniquely suited to solve this problem?" or "What is your unfair advantage?" Consider proactively steering the dialogue in this direction.

- Know your competitors, how you stack up against them, and how you'll disrupt the status quo.
  a. Don't tell an investor that you have no competitors. If it's true, you may not be pursuing a very interesting problem.
  b. Even if you don't think you have any direct competitors today, remember switching costs and inertia are formidable challenges to overcome.
- Cater your message to your audience.
  a. Typically, investors do not have prior PhD-level experience in your start-up's specific field, so don't get too into the weeds on the technology during a first call unless expressly asked to do so.
  b. How your technology works is of course important, but initially, how your technology is solving a big urgent problem is more important.
- Explain why you believe the specific venture fund you're speaking to would be a great fit for your company. Do your homework and learn about the investor.
  a. Maybe there could be some synergy with one of their other portfolio companies or perhaps one of the LP relationships could be valuable.
  b. If their investment focus or thesis is around something beyond financial returns (e.g. climate change, job creation, supporting founders from diverse backgrounds, etc.), present your company in the correct light.

12.5.1 Summary

When considering financing options, you need to examine the trade-offs and ask yourself what you're willing to accept. Your initial financing decisions can have a long-term impact on your company's trajectory. At the same time, the financing options you have at an early stage (i.e. pre-revenue) are more limited compared to revenue-generating businesses. Knowing your options as well as the right questions to ask will put you in a stronger position to make the best decision for your company and position it for success. The following is a quick checklist of some questions that can be handy to keep in mind for any fundraise. Remember, there is no such thing as a free lunch.

- How much risk am I personally willing to assume?
- What is the stage of my company, and what funding sources are available to me now?
- What is my end goal? Do I value maintaining ownership, a quick exit, or growing huge?
- How much time am I able or willing to dedicate to fundraising?
- Do I need money for short-term needs or to fund long-term growth?
- Am I willing to give up control of my company to bring in outside capital?
- How much dilution am I willing to accept?
- Can the funder provide value beyond simply money?
- Is taking this money putting me on an unattractive or untenable growth path?
- Are there any financial covenants to worry about?
- What are the expectations of the investor with regard to exit or payback?

There are no right or wrong answers for the preceding questions or for how to fund your commercialization effort. I hope this chapter has given you some insight into navigating the investment landscape. The final section of this chapter describes the funding journey of a sustainable chemical technology success story.

12.6 Financing Case Study: Cnano Technology

Cnano Technology is one of the world's largest producers of multiwalled carbon nanotubes (MWCNTs). It has successfully formulated its nanotubes into pastes that are integrated into lithium-ion batteries to improve energy density, power density, safety, and cycle life. The company leveraged a range of funding sources to support its growth into a profitable, publicly traded, advanced materials unicorn (i.e. valuation over $1 billion).

Cnano's MWCNT production technology was developed and spun out from Tsinghua University in 2007 with $6 million in Series A funding from a syndicate of VC investors including Pangaea Ventures, CMEA Capital (now called Presidio Partners), and WI Harper.

Pangaea was convinced that Cnano had an elegant method to make CNTs that was lower cost and more scalable than the competition. However, it wasn't a rocket ship to a billion-dollar company overnight. A better technology to produce a value-added material or chemical is a nice place to start, but it isn't often sufficient to become a viable business. Remember, investors eschew a technology looking for a problem. Determining product-market fit and solving an urgent market problem are still essential to achieve success. For Cnano, this meant formulating its MWCNTs into conductive pastes that could be used in existing lithium-ion battery manufacturing lines while conferring significant performance benefits. With the strong tailwinds of a burgeoning battery market thirsty for better technologies enabled by better materials, Cnano had the right product offerings at the right time to scale its top line (revenue) and bottom line (profit).

Note that this neat and clean success story glosses over some of the ups and downs and false starts encountered along the way. Initially the company wasn't focused on the energy storage market. Instead, they saw promise in improving fiber-reinforced polymer composites with nano-additives. Ultimately, this has become another successful area of their business, but it took many years longer for that market to mature and for Cnano to gain meaningful traction. The lesson here is that start-ups can and need to adapt and shouldn't be too wedded to their initial plan. In the famous words of boxer "Iron" Mike Tyson: "Everybody has a plan until they get punched in the mouth."

After the Series A funding round, there was a $7 million Series B in 2011 led by IDG Capital, a $16 million Series C in 2015 led by GRC SinoGreen Fund, a $48 million Series D in 2017 led by CICC, and finally a $5.7 million Pre-IPO Series E round. This culminated in a $130 million capital raise via an IPO on the Shanghai Stock Exchange in September 2019. The market capitalization on the first day of trading was an impressive $1.5 billion, and as of June 2020, it's currently trading higher than the IPO pricing (https://markets .businessinsider.com/stocks/jiangsu_cnano_technology_a-stock).

Beyond the equity funding to fuel Cnano's growth into a $1 billion+ business, they also leveraged Chinese government grants and provincial loan programs to support construction of their manufacturing facility. This amounted to an additional ~$5.5 million in funding. Furthermore, the company benefitted from a paid strategic commercialization partnership bringing in some additional financial support. Once the company reached profitability, they were also able to reinvest profits into the business as needed. Cnano essentially utilized financing from all of the major sources mentioned in this chapter to great effect – grants, strategic partnerships, equity investment, debt, and bootstrapping.

As previously discussed, investment comes in many forms and will vary a lot depending on the market, business model, and maturity level of a start-up. There isn't one correct way to finance commercialization, but I hope this chapter has shed some light on the investment landscape and provided some insight on how to navigate it.

Reference

1 Rafael, C.V. (2013). *The Future of the Chemical Industry by 2050*. Weinheim, Germany: Wiley-VCH Verlag GmbH & Co. KGaA.

13

Operationalizing a Start-Up Company

Andrew White

CHAR Technologies Ltd., Toronto, ON, Canada

13.1   Introduction

The operationalizing of a start-up company has two fundamental aspects – systems and oversight. Systems are the formal by-laws, policies, and procedures that describe how people in the company should actually get things done. They are critical in all aspects of a start-up, allowing for operation but also rapid growth. Systems are at the heart of sales, product development, project management, human resources (HR), health and safety, and financial management.

The other aspect is oversight. Oversight involves ensuring proper strategic direction of the company is being maintained and that the correct systems are in place. Oversight can be in the form of an advisory board who can provide independent advice or a board of directors who not only provide advice but also have a legal and fiduciary responsibility to the shareholders.

Having both systems and oversight in place in the early stages of your company will not only help with overall growth, but will make that next investment round much easier for everyone involved.

This chapter will present information on setting up oversight boards, both advisory and governance. This will be followed by an overview of key systems required in a start-up, including HR, health and safety, and finance with an additional look at how to use financial systems to make projections.

The information in this chapter was gained through operationalizing my own company, CHAR Technologies. As a graduate student at the University of Toronto in the Department of Chemical Engineering and Applied Chemistry, my thesis evolved from the initial research question to a patentable discovery. We (my thesis advisor Prof. Don Kirk, our lab's associate researcher Dr. John Graydon, and myself) developed a patentable process – the conversion of low-value fibers, derived from the anaerobic digestion of organic waste, into higher-value activated carbon. After working in the Engineering Department at a large provincial utility, I knew I wanted the opposite of that career path, and a start-up based on the technology I helped develop was a very exciting prospect. After finishing my MASc, I enrolled in a masters of business, entrepreneurship, and technology (MBET) program at the

How to Commercialize Chemical Technologies for a Sustainable Future, First Edition.
Edited by Timothy J. Clark and Andrew S. Pasternak.

University of Waterloo. It was during this program, in February 2011, that I founded CHAR Technologies. Since that time, we have completed a number of financing rounds, which we leveraged with grant programs specifically designed to advance the technology from lab to commercial scale. To facilitate financing, we brought CHAR public on the Toronto Venture Exchange and acquired an engineering services and technology firm. We now have a commercial-scale facility based on the same process I worked on at the lab scale during my MASc all those years ago. Most of the information presented here was not formally taught to me. Although I acquired a great foundation from my MBET degree, much of what I know I had to learn on my own as CHAR progressed from the lab to full-scale operations and being publicly traded on the stock market.

13.2   Oversight Boards

The most important distinction between an advisory board and a board of directors is responsibility. An advisory board can be made up of experts in business, your specific technology, or even start-ups in general, but they have limited responsibility to you or the start-up. A board of directors, on the other hand, has a real legal and fiduciary responsibility to the shareholders, and they are therefore open to legal action if anything at the start-up goes sideways. As the name might indicate, an advisory board is an important source of advice, whereas a board of directors' primary function is governance. Most investors, private or public, require a board of directors to protect their investment and make sure the company moves in the right direction.

There is one significant similarity between an advisory board and a board of directors. Both types of boards should be made up of seasoned and experienced individuals who can be tapped to provide guidance, advice, and, most importantly, access to their networks. Access to these networks will be critical to help find future employees, potential investors, strategic partners, prospective clients, etc.

13.2.1   Advisory Board

An advisory board can serve a range of purposes across the company lifecycle, but the overarching function is to be a sounding board for management and a source of independent advice. In the early stages of a start-up, the management team is likely lean, and individuals are wearing multiple hats (the classic dual role of CEO and CBW – chief executive officer and chief bottle washer). Having access to independent advice is critical to a start-up to help navigate through major decisions, avoid pitfalls, and prevent decision paralysis. During times when decisions seem so daunting and you don't know where to start, having a good advisory board is a great advantage.

Since you'll be wanting to build up your advisory board with experts, there is another major benefit – they provide a network. Building a robust network early on will be an absolutely critical success factor, and through the advisory board you can tap into the members' networks. These networks can be critical in finding funding, clients, strategic partners, future hires, and everything and anything!

The number-one, most important aspect of an advisory board is *you*. The term to use is "coachable." If you are not coachable, you are not enabling your advisory board to provide advice – and since an advisory board won't necessarily have skin in the game, there is little incentive for them to provide that advice. Being coachable doesn't necessarily mean you simply do what your advisors say, but it does mean you park your ego and be open to constructive criticisms. Wearing multiple hats inherently means you'll make mistakes, and the advisory board can help ensure these mistakes are minor, but only if you let them.

13.2.2 Board of Directors

A board of directors is distinct from an advisory board. As a starting point, in most jurisdictions, directors are a required part of the articles of incorporation. Although at the very early stages, the directors could simply be the founders, this is definitely not proper governance. The board of directors fundamentally serves two key functions. First, they provide oversight of the strategic direction for the company. Second, they are responsible for monitoring company activities and making sure that the strategy is effectively being executed.

Directors must maintain a fiduciary responsibility to act in the best interest of the company. This applies to directors representing shareholders as well; if a potential conflict prevents them from doing what is best for the company, they must declare it and ensure that the company's interests come first.

Having the right directors involved early on and enabling the board's structure to grow with the company will be critical to ensure shareholder value. This in turn will be favorable for the CEO/founder (as a major shareholder) and will help give confidence to potential investors.

As my board chair likes to put it, a board should conduct itself with "noses in, fingers out." That is, a board should know what's going on and provide oversight ("noses in"), but in general not be responsible for day-to-day management ("fingers out"). The board really has only one direct report, the CEO, and it is their responsibility to ensure that the CEO is effectively running the company. In the early stages, the lines may blur between management and the board, as often they will be the same. However, as the company grows, the structure should be set up such that the board is scalable and can begin to showcase effective overall governance. The easiest way to ensure the structure is in place for scalability is to have the option of a number of board members in the articles of incorporation. The language would be along the lines of "…a minimum of x and a maximum of y directors." By setting x at 1 and y at 10, there is more than sufficient room to continually evolve the board with the company's growth.

A properly constituted board will also have committees, with the audit and compensation committees (the latter is sometimes included in an HR committee) being regarded as the most critical. The audit committee will take responsibility for reviewing the financial statements as prepared and ensuring they are confident in the numbers as presented. The compensation committee will help determine the CEO's salary and bonus package. While both committees are not necessary in the beginning stages of a start-up, in particular if there is only one director, understanding what makes up a strong board will help in the early decision-making when appointing directors and building your board.

Boards are usually comprised of an odd number of members, which allows the board chair to cast deciding votes. In the early stages, a board made up of three members is more than adequate. Of the three, there should be at least one independent director, who is a person with no contractual connection to the company or its shareholders and can therefore act in a completely unbiased manner. The other seats could be for the founder/CEO and possibly an early-stage investor.

It is important to ensure the independence of the board as they are responsible for company oversight. If management holds all of the board seats, there is a potential conflict of interest, and they are not likely to provide adequate oversight. For example, they may be conflicted about their compensation and bonus or decide to take the company in a direction in which they have vested interests but doesn't represent good long-term value for the company and other shareholders. As the company grows, more seats can be added, and it's usual to keep the number of seats as an odd number. As the CEO/founder, an important consideration to bear in mind is that each new board member is an additional person that needs to be managed. It's not typical to see boards much larger than 5 for companies that are generating less than $100 million in revenue. As the board increases in size from 3 to 5, it is strongly recommended that the two new directors are independent. Since a board should be providing objective oversight, having the majority of the directors be independent will help ensure that's the case, ultimately bringing value to the shareholders.

13.2.3 Building an Advisory Board

In building your advisory board, you must first determine what skill sets you need for them to be effective in their roles. These skill sets can evolve over time as the company grows and changes, which means your board must also evolve over time. Once the positions are defined, you can then conduct a search and get the right people in the right chairs.

For the advisory board, you can choose just one member if that's what you need at first. If you are just starting out and your primary background is in the technology itself, then there are lots of different skill sets you should seek out. In the early stages, being able to tap into experience to help guide the business model should be at the top of the list. For example, if your company is developing a new battery chemistry, having an advisory board of experts in this industry will help guide the early decisions. Do you make the new battery system yourself? Do you find a strategic partner to co-develop it? Do you simply license the technology, and if so, how far along does the development need to be? These questions become easier to answer with an advisory board – they shouldn't be answering the question for you, but helping you understand the ramifications of each direction so that you can make a well-informed decision.

Often, once a start-up has grown into a small-medium enterprise (SME), the make-up of the advisory board will naturally change as well. Ideally, with most functional areas of the business systematized, most advisory boards move to providing advice in two fundamental areas – overall strategy and technology development. If a technology was invented in a university lab, sometimes the principal investigator and/or co-inventors are not interested in being part of the resulting start-up. Instead, having them participate as longer-term members of an advisory board can help keep them engaged with the start-up and its evolution. If they are inventors of the original intellectual property (IP), they likely need to be involved anyway if IP rights negotiations are taking place with the university as part of the start-up's formation.

13.2.4 Building a Board of Directors

Building a board of directors can be a daunting task and may seem less important than other tasks. However, providing for proper governance early on will allow the board to grow easily with the company. As with advisory boards, the first step is to determine which skill sets and functions are required. You will almost certainly need finance and HR expertise. Beyond that, the choices you make will depend on your technology and how you envision the business growing.

As mentioned, the early make-up of the board will be quite small. When searching for a director, a key (and difficult) mindset for the founder/CEO is to look for a director who might fire them. It's critical that the founder/CEO looks for a director from the shareholder perspective – after all, the founder/CEO is a major shareholder, so the biggest concern should be the survival and growth of the company.

At a high level, what makes a good director is experience and commitment. Experience should be relevant for the current size of the company. Is the potential director's experience from being on the board of a billion-dollar public company, or do they have experience as a director of a private start-up? Do they have either direct or translatable experience in your industry? Commitment is important. As a board member, they are not expected to commit significant amounts of time, but one day a month is not an unreasonable request. Will they be able to respond to requests urgently? Will they be willing to attend board meetings in person? Furthermore, a board also requires management from the CEO. Therefore, does the founder/CEO have a trusting and respectful relationship with the prospective director?

13.2.5 Managing the Board

Once the board has been established, it will need to be managed on an ongoing basis. Managing a board is a different task than managing employees. In between official meetings, directors may want differing levels of detail reported over different intervals (e.g. once a month versus once a week). Getting nonstop requests from directors is a common challenge and must be managed for you and your business to remain productive and efficient. Proactively managing a board by providing regular written updates will help the founder/CEO from spending too much time responding to ad hoc requests from various directors. Providing regular updates will also help keep the directors abreast of the development of the company.

This is important not only from a board management perspective but also to help maximize the potential value (advice and network) that directors bring. It will also help board meetings to be more efficient and productive, as the directors will be "in the loop," and precious meeting time does not have to be spent getting people up to speed.

Additionally, much like with every aspect of the business, systems should be in place to monitor and evaluate the board's performance. There could be a potential conflict inherent in the CEO evaluating the board, since the board has the capacity to fire the CEO. Thus, the CEO should be responsible for ensuring board performance reviews are a priority. The board should be responsible for conducting a self-assessment, ideally with metrics that have been tracked to bring in a level of objectivity. If there are performance issues at the director level, they should be dealt with accordingly. As a CEO, every team you interact with should be filled with A players (versus B or C players), and the board should be no different.

13.2.6 Compensating Boards

In general, remuneration should be relative to responsibility. An advisory board doesn't have responsibility of any real significance, but you do want to keep them engaged. An important consideration is the idea of an honorarium – a payment made for services provided in a voluntary capacity. The honorarium concept should help with deciding on a magnitude of remuneration. As a start-up, cash conservation must be a critical consideration. Additionally, as a start-up, equity (shares) may seem like a "cheap" form of compensation, but are actually quite expensive.

One way to not only compensate the advisory board, but also keep them engaged, is to issue them stock options. Stock options are exactly that, an option to buy a set number of shares in the company at a set price. This set price doesn't change as the stock (ideally) increases in value over time. The incentive is that as the value of the company increases, the share price increases until they are worth more than the option price. At this point, options are "in the money." The option owner can buy the shares at the set price and elect to sell them at the market price if they want, making a profit in the process. For example, if the option is set at $0.05/share, and the company's market value would dictate a price of $0.20/share, the optionee can still buy shares at $0.05 and sell them to make a profit of $0.15 per share.

If the company remains private, a key aspect of the stock option agreement will be to allow optionees the ability to buy the options and sell them to new investors when the company is raising funds. Alternatively, a mechanism can be set up that compels the company to purchase the shares once the optionee has converted their options to shares. This mechanism should be carefully considered, as cash conservation will be critical in the early stages, and having a mechanism whereby the company may be obligated to buy back shares from an optionee is inherently risky.

While there is no formula for setting option prices, here are some standard considerations:

- **Number of options granted**: Obviously, providing more shares results in greater incentive for the owner. This can be staggered over time as well. For example, an advisor could be granted 1000 shares per year for each of the first three years served.
- **Price**: This is always difficult to determine. If the start-up is so new it hasn't yet raised money, a good way to manage the price is to delay setting it until a share price is established once an initial financing round is undertaken.
- **Vesting**: This is when the granted options can actually be converted and cashed in. A time can be set for vesting, such as immediately or delayed until the advisor fulfills specific requirements. For example, after one year of being an advisor or the start-up completing certain milestone activities.

An advisory board can be a low-cost way for a start-up to access robust independent advice and input, which will help make sure those inevitable mistakes are relatively minor in nature.

Compensation will be more significant for the board of directors than for an advisory board, as they have greater oversight, including a legal fiduciary responsibility. Again, as CEO of a start-up, cash conservation should be of paramount importance, and therefore it's recommended that compensation be equity-based. Stock options have advantages over straight equity (including beneficial tax implications for the directors). Vesting should be considered as well, with options vesting annually being a more standard practice. As a rule of thumb, anywhere between 0.25% and 1% of the company's overall shares (in the form of options) per year of service is a good place to start. Once the company has achieved a significant milestone in terms of revenue or investment, a cash honorarium can be added on a per meeting basis (for example, $1000 per meeting attended). While in the early stages, directors should be encouraged to pay their own way for travel, once cash is more available, it is standard to cover directors' travel.

13.3 Systems

As alluded to at the beginning of this chapter, the success of a start-up will come down to many factors – having robust systems in place from the outset will best position a start-up for success. Every part of a start-up can (and should) be systematized. Most systems are well developed and broadly available, be they in sales, product development, project management, or financial in nature, among others. Leverage the expertise of your advisory board members who have been in the trenches to set up all these systems. It doesn't mean you can't customize the systems for your start-up, but what it does mean is you shouldn't need to start from scratch and re-invent them.

13.3.1 Human Resources Management

In many ways, HR management can be the most significant challenge. Assuming the initial founding team is in place, the next step for growth has a number of options. If there is a discrete scale-up project, do you hire someone permanent/full-time or on contract? Or do you bring an experienced consultant into the fold as recommended in Chapter 11? How do you make sure the potential hire is the right fit? How do you ensure that the core competencies of the start-up are maintained and enhanced? While there are no right answers, this section will endeavor to provide a framework to evaluate these decisions.

For a start-up, HR is invariably linked to financial resources. There are always too many things to do, but how can we afford to hire someone to help? If you find your first thought when facing the resource constraint challenge is "I need a CFO/COO/market manager, etc.," you've already skipped some critical steps in ensuring that when you bring on additional support, you're finding the right person or people.

To start with, what are the discrete tasks that need to be accomplished? The answer shouldn't be "we need a marketing plan." Instead, what tasks go into creating a marketing plan as described in Chapter 2? Understanding the competitive landscape? Understanding customer needs? Understanding your distinctive competencies? Also, what kind of data, processes, or tasks need to be accomplished to answer these questions? If the answer is "I don't know," then the task list should be more along the lines of developing the framework for a marketing plan rather than developing the marketing plan itself.

With a task list in place, some fundamental questions are now:

 i) Is this a discrete project, or will it be ongoing?
 ii) Is this a core competency for us?

If it's a discrete project, you may consider hiring a consultant. A good consultant can bring a wealth of knowledge and experience to help quickly accomplish a project. There are two angles to the core competency question. First, is this something we already know how to do but don't have the human capital to complete it? If this is the case, perhaps a contract employee is the right approach as it is something the start-up knows how to do. The other angle is, is this a task that is important to the development of our core competencies? Consultants can help, but you need to consider how to transfer the knowledge in-house that has been developed.

Consultants can play a key role. They usually have significant experience in their field that they can bring to the start-up. Their fees can be budgeted, remuneration can be negotiated to be contingent on successful outcomes, and they are no longer an expense item after the project is completed. Hiring, on the other hand, helps develop the start-up's capabilities for the long term, but it is important to remember that an employee is a constant cash burn, whether there is an active project or not.

If the task list development has led to the decision to hire someone, the next step is to develop a roles and responsibilities document for the position. This process will help clarify if this will be a newly created full-time position or whether some of the roles and responsibilities can be shared between your current employees.

A key consideration regarding staffing is the start-up's culture. To be clear, culture is *not* having a few ping-pong tables around the office (although ping-pong tables can support the culture). Culture can be defined as a set of shared values, goals, attitudes, and practices that characterize your company. At the outset, determining the defining aspects of your culture can be accomplished by *honestly* working through your company's mission, vision, and values. "Honestly" means spending the time to really work together as a founding team to determine these three overarching cultural aspects, not simply choosing some key words. As a founder, it's also important to determine and demonstrate how the values dictate how you work with each other, with clients, and with vendors. Culture should also address the big "why" questions. Why do you exist? Why are you doing what you are doing? Why should anyone (stakeholders, shareholders, employees, clients, etc.) care about what you do? If you can't express your company's culture, then you should not be trying to hire any employees. Understanding your culture and hiring for cultural fit are critical aspects for long-term success. If there are tasks that are time sensitive and your start-up doesn't have a defined culture, this is a good indicator that a consultant may be a better fit until you set your culture.

This section won't go into further details of company culture, except to point out that when done wrong (or, more likely, not done) the start-up's chances of success are drastically reduced. When done right, there's nothing that the team can't accomplish. It's the framework and understanding of how everyone conducts themselves in the organization, and while it may seem unimportant, it is a critical factor in the success of the start-up.

If it's been determined that a new employee is needed, the most important (and true) business adage is "hire slow, fire fast." Ideally, if you get the first part right, you don't need to worry about the second part. If undertaking the hiring process internally, there are a number of tools available. At a basic level, reading some details on how to conduct a behavioral-based interview is important. You may have experienced this before as they usually start with "tell me about a time...." These questions should be designed to help assess fit for both roles and responsibilities as well as fit with the start-up's culture. Take as much time as needed to make sure that the candidate will be the right fit. Have three, four, or more interviews and make sure various members of the team, not just management, interview the candidate. Making the wrong hire can be extremely expensive, not only from a salary and severance (if they need to be exited) perspective, but also on the productivity and engagement of the rest of your team.

Finally, it is well worth the investment to hire appropriate legal counsel to help draft an employment agreement. If the employee needs to be exited at some point, a strong employment agreement with well-defined terms will make an already uncomfortable situation that much easier on everyone.

13.3.2   Health and Safety Systems

A strong health and safety (H&S) system and focus is paramount for any company of any size. H&S regulations are local, regional, and national in nature, so it is first important to know the law. It also means that each jurisdiction that the start-up may operate in will have their own mandated employee health and safety requirements. Following these rules and regulations is critical but is also the relatively easy step. A pertinent business adage is "leaders get the behaviors they exhibit or tolerate." If the H&S policies and procedures your company has implemented are routinely ignored by employees without repercussion, or management themselves ignore them, it's unlikely that your team will feel a strong commitment to H&S. As a result of this, they may put themselves in harm's way believing that the culture allows for it. This is unforgivable.

As with all cultural aspects, an unwavering commitment to H&S needs to be demonstrated by management and enforced when necessary. A good place to start is by codifying this commitment in the form of an occupational health and safety policy. Then begin working on a program to implement it. There are many elements to a comprehensive H&S system, with the following list providing starting points from which one can be built:

- **Joint health and safety committee (JHSC)**: A JHSC is a requirement in most jurisdictions and is to be made up of employees and management. Various jurisdictions will have prescribed requirements (e.g. monthly inspections of fire extinguishers and emergency exits). You should view the JHSC as an opportunity to continually promote the company's culture and absolute commitment to safety, not as a bureaucratic hurdle. It is important to remember that most H&S requirements and regulations have been implemented *after* a major incident causing injury or death. There is a very real, and very serious, reason for their existence. You should empower the JHSC to monitor, observe, and raise concerns beyond the monthly requirements and to promote safety among the entire organization. It is also critical that when the JHSC observes and raises areas of hazard or risk that these issues are actioned on, not simply documented and left as is. An important cultural implication is one of ensuring safety, not just of "doing one's job" by documenting the risk.
- **Hazards analysis and risk assessment**: Hazards analysis and risk assessment is focused on the prevention of work-related incidents. Any project work should be initiated only when it can be completed safely, with a high degree of confidence that all hazards have been identified and addressed and that any remaining risks are manageable when knowledgeable, trained people perform the work. The JHSC should be involved in reviewing the analysis. Once hazards are identified, what-if scenarios can be used to help address any risks. For example, if working with a pressurized system, is it possible for the system to become overpressurized? What controls can be put in place to reduce the risk or likelihood of overpressurization? If these controls fail, is there a redundancy (i.e. pressure relief valve)? If the pressure relief valve activates, are the resulting emissions hazardous, and if so, how can the risks related to the release be mitigated? As a real example, I developed our activated carbon material primarily to remove hydrogen sulfide from biogas. Hydrogen sulfide is a toxic, corrosive, and dangerous substance. We were blending H_2S with a carrier gas, which was then passed through an adsorption column to determine how much H_2S was adsorbed on the various biocarbons in the column.

Some of the process modifications we implemented to reduce various risks included (but were not limited to) the following:

i. Using a custom blended gas cylinder of 1% H_2S in N_2, instead of a cylinder of pure H_2S

ii. Ensuring any tubing connections were solely within the fume hood

iii. Placing an H_2S monitor adjacent to the gas cylinder to measure any potential leaks from the cylinder regulator

iv. Ensuring the effluent gas used for testing passed from the adsorption column to secondary treatment

v. Ensuring others working in the vicinity, even if not working on the specific project, were aware of the use of H_2S and safety protocols

vi. Preparing and posting very clear emergency instructions that left no room for interpretation (unclear language can lead to hesitation in an emergency)

- **Incident reporting and investigation**: Report, report, report. Relatively minor incidents can portend major incidents. Treat incidents like valuable information, not as a bureaucratic pain – these are opportunities to learn and improve procedures in a proactive manner. Complete a thorough root-cause analysis, and continue to dig as deep as possible by evaluating and asking why until you arrive at the root cause(s). To reiterate, incident reports should be treated as a source of information, not as a means to ascribe blame. Employees should feel free to report any incidents, without fear of reprisals and with the understanding that management takes all incidents and investigations seriously. Incident investigations must also lead to the implementation of any corrective actions, including changes to procedures or engineering controls. These should be completed in the spirit of continuous improvement, which will also help reaffirm the company's culture.

- **Training**: As with incident reporting, training should be viewed as an opportunity for continuous improvement. Any new technical work (new to the company or new to the employee) should include specialized training related to that work. Any technical work should also include periodic refresher training. At a bare minimum, WHMIS training and refreshers should be codified into the H&S program.

While all businesses deal in various levels of risk and start-ups are seen as particularly risky endeavors, it's critical to understand that there is an additional layer of risk with chemistry/chemical engineering–based start-ups. The ultimate accountability for the health and safety of everyone involved in the start-up starts and ends with the founders and management. A serious H&S incident or infraction could spell the end of the start-up. Enormous fines and potential jail time are real possibilities if there is a preventable incident. Far worse is an incident may mean a permanent impact to, or end of, someone's life. Making your H&S system front and center at your start-up will ensure the focus is on everyone being able to go home safely at the end of the day and will help reaffirm a strong culture where employees know they are valued members of the team.

13.3.3 Financial Systems

Setting up proper financial systems from the start is a critical step in operationalizing any start-up. To the founder of a sustainable chemistry start-up, financial statements, forecasting, reporting, accounting, and bookkeeping can all seem like daunting tasks. This is especially true if at some point in the future you need to work backward and catch up on creating all the required financial statements. For example, a potential investor may want to see financial statements for the last few years, which is an unwelcome surprise if you are not prepared. However, if a working financial system is set up correctly from the start, providing an up-to-date financial statement to a potential investor (or bank or partner or government support program) can be quick and easy.

For someone with a STEM background, setting up appropriate financial systems is all about developing and maintaining a robust and reliable dataset. That all starts with bookkeeping – keeping a record of your business' financial transactions including purchases, sales, receipts, etc. The most important factor in managing financial records is to make sure you just do it! Furthermore, it can be done very inexpensively by the start-up team in the beginning. There are a number of reputable online programs available, some of which are even free with a small enough number of entries. You also don't need to choose the most appropriate program right away. A simple program like Wave (designed with a small business owner in mind) can be sufficient at the beginning. As long as the chart of accounts can be exported into a Microsoft Excel or CSV format, the data can ultimately be used with more complicated programs that may be implemented as the business grows.

One key aspect of bookkeeping software is its ability to create backups and proof of transactions. For example, if you've made a purchase of any size, you need be able to upload and link the invoice and receipt with that particular transaction. This will be important if and when the start-up scales to the point where investors and/or collaborators want to see audited financial statements.

If it still seems daunting after reviewing bookkeeping options, there are many firms that can manage bookkeeping for you. While slightly more expensive than using an online program yourself, it is likely that the outsourced bookkeeper will be using a more robust and recognized program (like QuickBooks) that can scale relatively easily with you. However, much like bookkeeping internally using software, it is critical to continually send data to the bookkeeper so the financials can be constantly updated. This will allow the financials produced to accurately reflect the status of the start-up at any given time.

Systematizing the bookkeeping is also an important first step for integrating proper financial controls. If the cash flow statement isn't corresponding with the bank statement, it's much easier to run a forensic review to determine whether a purchase or investment wasn't recorded properly or whether there may be an instance of fraud. Although the idea of fraud is a pretty negative thought among the excitement of a new start-up, it can happen. This risk also increases during rapid scale-up when hiring many new people.

Once bookkeeping becomes second nature (5 to 30 minutes at the end of each day was my routine at the beginning), all the other systems are much easier to work with. Financial statements are a good way to check on how the business is doing on a given day. The three most used statements are the following:

1) **Balance sheet**: This document provides a snapshot of the financial state of the business at that moment in time. It summarizes the company's assets, liabilities, and shareholder equity.

2) **Profit and loss (also called the P&L or income statement)**: This shows how the company performed during a period of time and includes sales, accounts receivable (amounts owed to the company), and other transactions.
3) **Cash flow statement**: This document also shows how the company performed during a period, but with a focus on actual cash in the bank.

Without getting into too much detail, it is important that the profit and loss and cash flow statements be read together. Often the cash flow statement is overlooked, but it is especially critical. Your P&L could show great profit, but if you're also investing significantly in capital equipment for scale-up, you may find yourself short of cash. Furthermore, sales do not necessarily mean you have been paid. Some companies will pay you 60 days later, but the P&L will still show that the sale is complete! The money that you need to purchase that equipment may not yet be in the bank. This is why following the cash flow statement is key because at the end of the day, cash is still king.

Additional financial report stakeholders are government funders and support programs. Different programs have different rules. Here are some examples:

i) **Stacking**: Even if the specific program is providing only a percentage of the project funding, they may stipulate that total project funding can be supported only up to a certain percentage by all government funding sources.
ii) **Expense claims**: Some programs are set up on a reimbursement process, which means that you need to spend the money and then submit an expense claim (with proof) to receive the funding from the program.
iii) **Final reports**: Inevitably, government support programs will ask for a final report that will require a financial component.

These are just some examples where reporting or providing evidence is significantly easier with a proper bookkeeping process in place.

Early investors, partners, and banks will also want to see your financial statements, which with proper bookkeeping systems in place will be relatively easy to provide. As the start-up grows, these interested parties may ask for "notice to reader" statements or "audited" statements. Both of these statements are prepared by accounting firms, and both will be much easier (and therefore much cheaper) to provide with strong internal bookkeeping. Notice to reader statements are reviewed by a firm as a quality check on the reporting. Does the balance sheet balance? Does the cash flow, P&L, and balance sheet all integrate together? A notice to reader does not check the quality of the underlying data. For that, an accounting firm will prepare audited statements, which will include looking for evidence to support the data. For example, if there is a capital purchase or expense of $50 000, is there an invoice to prove that the amount was paid? With robust bookkeeping and all proof of transactions routinely uploaded and linked to each transaction, the audit process will be much more manageable and cost effective.

13.3.4 Financial Projections

Financial statements will either show the start-up at a moment in time (balance sheet) or its performance over a period of time (P&L, cash flow). However, most potential stakeholders want to see what the future might look like. This is where projections come into play.

Financial projections need to be both optimistic and realistic. If the projections show the company ramping up from $0 revenue to $100 million in revenue over a very short time frame, they should also show realistic investment requirements to get there.

Fundamentally, projections should show potential investors or other stakeholders the following:

i) How much revenue will the company earn in a given period of time (usually years)?

ii) How much revenue will be needed to get to a break-even point? This is the point where revenue covers expenses.

iii) How much will it cost to get to that revenue target? More importantly, will this investment round be sufficient, or will future investment rounds also be necessary?

iv) Indirectly, what's in it for the investor? How do they make money?

Investors *may* ask for projected balance sheets, P&Ls, and cash flow statements. However, I've found success in a combined P&L/cash flow hybrid approach. Balancing projections that include accounts receivable (AR) and accounts payable (AP) assumptions is tedious in my opinion. The timing of cash is of *critical* importance, and this is of course influenced by AR and AP. There is a simplistic way to address this in a hybrid projection that won't require hours of troubleshooting a linked Excel document. Here's how you set one up – see Figure 13.1 for reference.

The top section will be the P&L, and the bottom section will be a simplified cash flow statement. The P&L statement portion can be copied from your bookkeeping/accounting software to make sure it will link correctly with your accounts. The major rows will be revenue, cost of goods sold, gross profit, expenses (include noncash expenses such as depreciation), tax, and finally a net income number.

The simplified cash flow statement begins below net income. The three main categories (rows in your spreadsheet) of a cash flow statement are operations, investing, and financing.

For the operations section, you start with the net income and add back in noncash expenses (e.g. depreciation, amortization). The operations section is also where you can add a line for working capital, including a contingency for covering the cash timing of AR versus AP. The investing section includes any capital expenditures (e.g. purchasing a reactor or anything you spend cash on that is not captured in the expenses section of the P&L). The financing section will be cash into the start-up from investors or the bank, in the form of equity or debt. Note that it's good to have separate lines for equity and debt. If you are anticipating debt, a separate tab should be created to calculate interest (which needs to go into the expense section within the P&L section) and principal repayment (which can be recorded as a negative number in the financing section of the cash flow portion).

Net cash for the year is then calculated from the three subtotals: cash from operation, cash from investing, and cash from financing. Finally, take your starting cash and add or subtract net cash, to arrive at ending cash (which is, of course, your starting cash for the following year). If ending cash is negative, it means more investment will be required.

With this powerful exercise to arrive at a single projection, revenue, profitability, and required investment can all be seen in one place. Unless asked, I tend to keep my projections at three to five years out. Once the projections extend beyond five years, the uncertainty band for a start-up is significant.

	Month
Revenue	
Revenue	C6
Grant	C7
Cost of Revenue	C8
Gross Profit	**C9 = C6 + C7 − C8**
Expenses	
Interest	C12
Research and Development	C13
Professional Fees	C14
Office Expenses	C15
Regulatory and Filing Fees	C16
General & Administrative	C17
Depreciation & Amortization	C18
Total Expenses	**C19 = SUM(C12:C18)**
Net Income (Loss)	**C21 = C9 − C19**
Cash Flow Operations	
Net Income	**C24 = C21**
Allowance for AR/AP	C25
Depreciation & Amortization	**C26 = C18**
Net Cash from Operations	**C27 = C24 + C25 + C26**
Cash Flow Investment	
CapEx	C29
Cash Flow Financing	
Equity	C31
Debt	C32
Net Cash from Financing	**C33 = C31 + C32**
Net Cash	**C35 = C27 + C29 + C33**
Bank Balance, Beginning of Month	C36
Bank Balance, End of Month	**C37 = C35 + C36**

Figure 13.1   Sample hybrid P&L/cash flow approach for projections.

Depending on the skillset of the management team, board of directors, and board of advisors, the timing of bringing in a qualified "CFO" will be variable. I've put CFO in quotations to denote that it's additional financial expertise that may be required, not necessarily a titled CFO. As with many services, a fractional CFO is a good starting point when some expertise is needed, but a full-time CFO is not yet necessary. There are many options for fractional CFOs, and the best place to start looking is by tapping into your directors and advisors.

13.4   Conclusion

This chapter provided a high-level overview of oversight and systems required to get a start-up on the path to success. These topics are not covered in traditional chemistry or engineering programs, so they must be learned from others as you go forward. Take the time to engage with people who can advise and help you set up the company that you want. Even if they are not formally engaged, they can still provide a lot of value. A critical success factor for any start-up is the development of a robust network; the more engagement undertaken, the stronger the network becomes. Setting up appropriate systems takes time as well, and often trial and error is needed to make them effective. It's good practice to periodically evaluate how your systems are performing and to adapt and modify them as needed, particularly as your business evolves and scales. There is nothing wrong with making changes when you need to. In fact, it is expected from the effective entrepreneur and can be a micro-version of an important entrepreneurial skill, pivoting.

Part IV

Success Stories

14. Making an Impact: Sustainable Success Stories

14.1

CarbonCure

Jennifer Wagner and Sean Monkman

CarbonCure Technologies Inc., Dartmouth, NS, Canada

In 2007, CarbonCure set out to reduce the carbon footprint of the concrete industry. The company now finds itself at the forefront of a movement to turn CO_2 into this valuable commodity. By 2030, carbon utilization is expected to be a $1 trillion industry and reduce global greenhouse gas (GHG) emissions by up to 15%. According to the Global CO_2 Initiative (GCI) [1], CO_2 utilization products for the concrete sector alone can create an estimated $400 billion market opportunity and have the potential to reduce up to 1.4 gigatons of annual CO_2 emissions. While many companies are working in this space, CarbonCure has a substantial first-mover advantage as the only company with a market-ready, cost-effective, and proven operational CO_2-utilization technology.

14.1.1 The Vision

CarbonCure's vision is to make the incorporation of waste CO_2 into concrete the standard for all global concrete production. CO_2 utilization has been developed as a way to make stronger and greener concrete. Among nearly 200 developers of CO_2 utilization technologies evaluated, GCI ranked CarbonCure as the most scalable solution in the building materials segment to convert CO_2 into end products that are emissions-neutral or -negative. CarbonCure's portfolio of technologies has the potential to reduce up to 500 megatons of annual global CO_2 emissions and create up to $26 billion in new production efficiencies.

14.1.2 The Core of the Technologies

The core chemistry behind the firm's technologies were originally investigated by Rob Niven who, as a graduate student at McGill University, studied the benefits of adding CO_2 to fresh concrete. When CO_2 is added to concrete during mixing, the CO_2 creates carbonate ions. The carbonate ions then react with calcium ions released from the hydrating cement to rapidly form a limestone-like mineral, which develops as nanomaterials dispersed throughout the concrete (see Figure 14.1.1). This conversion of CO_2 into solid carbonate

How to Commercialize Chemical Technologies for a Sustainable Future, First Edition.
Edited by Timothy J. Clark and Andrew S. Pasternak.
© 2021 John Wiley & Sons Ltd. Published 2021 by John Wiley & Sons Ltd.

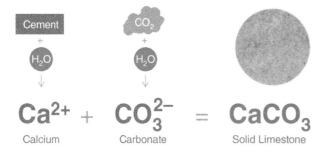

Figure 14.1.1 A schematic summary of the chemical reaction in the CO_2 utilization process that creates calcium carbonate ($CaCO_3$).

minerals finds it permanently bound within the cement matrix, and, as such, it cannot be released back into the atmosphere. The addition of CO_2 serves both to lower the carbon footprint of the concrete and to make the concrete less expensive to produce.

14.1.3   Determining the Value Proposition

The cement industry produces approximately 5–7% of global GHG emissions and faces increasing regulatory and public pressure about environmental impact. The cement and concrete industries are actively looking for solutions to reduce their carbon footprints. Environmental concerns and the vision to recycle cement plant CO_2 in the concrete industry led the company to consider CO_2 as an asset rather than a liability. Among the available solutions the cement industry identified, carbon sequestration and utilization represented the opportunity that could provide the most significant reductions.

Compared to competing technologies, CarbonCure's technology has several distinct advantages. The first is that the environmental footprint of the entire process is reduced without compromising the performance of the material itself. In fact, the resulting material is often stronger. Discussions with potential customers made it clear that not compromising on concrete performance was the foremost concern, with environmental benefits serving as a bonus provided the technology drove economic and/or business value.

The company determined that the addition of CO_2 could promote a strength enhancement, allowing users to use less cement, which greatly increases both the GHG emission reduction and cost savings to the producers. Technology adoption also followed a retrofit design that took advantage of lower costs of implementation and simple integration with existing concrete plants (see Figure 14.1.2). This allowed producers to adopt the technology through one small change while maintaining their existing equipment, materials, and expertise.

14.1.4   The Commercialization Pathway

CarbonCure was founded by Rob Niven in 2007. That year, Rob attended a United Nations summit on climate change, where he saw a global demand for solutions to reduce carbon

Figure 14.1.2 A CarbonCure system connected to a CO_2 tank at a ready-mix plant.

emissions. Inspired by the summit, Rob thought to himself, "The scientific community understands that CO_2 can be chemically converted to a mineral within concrete. So why can't we find a way to use CO_2 in everyday concrete and help concrete producers respond to the demand for green building products?" After that, he settled in Halifax, Nova Scotia, and recognized that he had to pursue an industrial validation of the research he conducted at McGill University by starting a business.

A management team was assembled that included board members with a history of cleantech innovation, cement and concrete industry experience, and intellectual property expertise. Strategic partners (an industrial gas company to supply the CO_2 and a precast concrete company to host industrial work) rounded out the consortium needed to take the idea from the lab to the industry floor.

14.1.5   Financing

In CarbonCure's early days, technology development was funded through consulting revenues from a sister company called Carbon Sense Solutions, which offered carbon management consulting to heavy emitters. This ability to raise cash by providing a service in the same industry was key to getting the technology off the ground. Additionally, some key grant programs supported the early research and development work. One of the first grants CarbonCure secured was through a program called Sustainable Development Technology Canada (SDTC). The initial SDTC-funded project aimed to refine and commercialize CarbonCure's first-generation CO_2-utilization technology in the concrete sector. To obtain this grant, CarbonCure had to demonstrate a pathway to develop the technology with a clear sustainable advantage over the incumbent technology. A technical validation plan, detailed

environmental analysis, and commercialization strategy were all part of the project proposal. CarbonCure had to demonstrate that there was demand from industry, creating a
viable business opportunity. SDTC was seeking technologies that could provide environmental *and* economic benefits – CarbonCure was the perfect fit.

The company was later able to obtain equity funding from a number of venture capital and angel investors through four rounds of financing. The attractive technology value
proposition, unique business model, and immense global opportunity were some of the
driving forces that led to buy-in from investors.

14.1.6   Development and Validation

In CarbonCure's early days, most of the effort was placed on understanding the fundamental science behind the carbon mineralization reaction and how to integrate the concept into
concrete production. By understanding how the science could intersect with manufacturing, the team of researchers and engineers could collaborate to design a system that allowed
for the precise delivery of CO_2 into concrete. Once the scientific barriers were overcome, it
became an engineering challenge to design a system that not only worked but also provided
a cost-effective solution to customers. Efforts were made to simplify the design, minimize
the equipment costs, and improve the return on investment while respecting the constraint
of a retrofit solution to industrial equipment and processes.

The first stages of technology development involved industry partnerships that were a
critical step into a real-world environment. The performance of the technology was validated through two full-scale industrial demonstrations with Halifax, Nova Scotia's Shaw
Brick and Brampton, Ontario's Atlas Block (now Brampton Brick) for the production of
concrete masonry units (concrete blocks). The partners were identified through their history of innovation and leadership within the industry and entered into the work in a spirit
of collaboration. These demonstrations allowed CarbonCure to progress its technology out
of the lab and to test it at an industrial scale in multiple full-scale production facilities while
gathering valuable experience and insights.

From 2009 to 2014, CarbonCure was able to use the in-market demonstrations to generate technical data and acquire customer feedback to improve its technology and value
proposition. Whereas the first iteration of the CO_2 injection step added the gas through the
mold cavity, adhered to a tight production cycle, and led to blocks with improved properties, the retrofit hardware was neither scalable nor sufficiently durable. A rethinking of the
approach settled on a new iteration where the CO_2 was injected earlier in the production
process; however, the hardware integration was not optimal. A final pivot led to CO_2 being
injected directly into the mixer. The hardware was scalable and durable while the concrete
met the desired benchmarks. A technology was ready for the market. Once the technology
for masonry products was refined, optimized, and commercialized, CarbonCure worked to
develop a solution for ready-mix concrete applications.

Through extensive lab work and collaboration with a local Halifax-based concrete
producer, Quality Concrete, the approach of incorporating CO_2 into ready-mix concrete

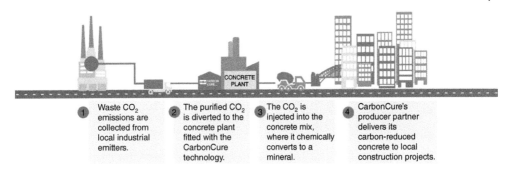

Figure 14.1.3 Schematic showing CarbonCure's end-to-end process.

evolved to use an optimal dose of CO_2 to maximize the performance benefit of the technology. The compressive strength increased without negative impacts on concrete workability. The technology team worked to refine the hardware, integrate with plant operations, and validate the value proposition and business model. By the end of 2015, the ready-mix commercialization effort had begun.

Ready-mix concrete production represents about 75% of the concrete industry; this application of the technology is where CarbonCure is seeing most of its growth today (see Figure 14.1.3). CO_2 injection methods have continually improved, becoming increasingly more effective and efficient as well as economical for concrete production. Not only is the technology cost competitive, but the end product can allow concrete producers to act sustainably while conducting smart business. Together with the overarching environmental benefits of the technology, CarbonCure has an attractive product that has garnered considerable global interest from strategic partners, customers, investors, and even Bill Gates [2].

14.1.7 Successes

To date, CarbonCure's technologies have been installed in more than 300 concrete plants globally. Partnerships with other companies from across the supply chain are being leveraged to service customers around the world. Over the first four years of commercialization, the concrete made with the CarbonCure Technology has surpassed 6.5 million m^3 and was used in high-profile projects including 725 Ponce de Leon Avenue in Atlanta, Georgia, and the McDonald's Flagship in Chicago, Illinois (see Figure 14.1.4). As of January 2021, cumulative carbon savings exceed 100 000 tons.

The current challenge facing CarbonCure is how to capitalize on the market opportunity in front of the company and maintain exponential customer growth. CarbonCure has quickly become the global leader in carbon utilization solutions. New innovations are under development to provide greater economic, logistical, environmental, and performance value to concrete producers. CarbonCure is transitioning from the simplicity of serving a handful of customers to becoming a mature and growing business looking to offer a technology, value proposition, and business model to more than 100 000 customers worldwide.

Project Details

Project Size: 360,000 sq ft
Architect: S9 Architecture
Developer: New City Properties
Engineer: Uzun + Case
General Contractors: Brasfield & Gorie
Completion: 2019
Concrete Supplier: Thomas Concrete
Total CO2 Saved: 680 tonnes

Project Details

Project Size: 19,000 sq ft
Owner: McDonald's Corporation
Architect: Ross Barney Architects
Completion: 2018
Concrete Supplier: Ozinga
Total CO2 Saved: 14 tonnes

Figure 14.1.4 Example of projects that leveraged CarbonCure's technology.

14.1.8 Lessons Learned

- Start simple. It does not need to be perfect; just get started. If you wait for the technology to be perfect, you may not get anywhere. Get started with early adopters who understand that the technology will evolve over time.
- Find the right partners. There will be ups and downs in the early days. Ensure that your partner is like-minded and open to trying several versions until you get it right. This takes time, and your early partners need to be patient.
- Listen to your customers. The beauty of demonstrating your technology with a real customer is that you get real operational data – do not take this for granted. Use this information to improve your technology.
- Be adaptable. Things will inevitably look different 10 years after your technology launch. Be open to feedback to develop the best version of your technology.
  Visit www.carboncure.com for more information.

References

1 CO_2 Sciences and Global CO_2 Initiative. (2016). Global roadmap for implementing CO_2 utilization. Report distributed by the Global CO_2 Initiative at the University of Michigan.

2 Gates, B. (2019). Buildings are bad for the climate. GatesNotes. https://www.gatesnotes.com/Energy/Buildings-are-good-for-people-and-bad-for-the-climate (accessed 4 November 2020).

14. Making an Impact: Sustainable Success Stories

14.2

Avantium

Gert-Jan M. Gruter and Thomas B. van Aken

Avantium, NV, Amsterdam, The Netherlands

14.2.1   Initial Technology and Business Model

Avantium was founded in 2000 with a technology that was developed in the 1990s by Royal Dutch Shell and transferred to Avantium as a shareholder contribution. The Shell technology was developed for accelerating catalysis research and development (R&D) by testing process catalysts in an automated, downscaled, and parallel fashion. In addition to Shell, several other companies invested during the launch of Avantium. These included Akzo Nobel, Eastman, W.R. Grace, GSE Systems, three Dutch technical universities, Pfizer, and GSK. Unfortunately, the first five years were quite turbulent as the diverse owners of the company were all interested in different technology directions. Avantium itself was also unclear in which direction to proceed and what business model to adopt. For example, should the company focus on catalysis for pharmaceutical process development, a robotic paint formulation and testing platform, or something else entirely? Also, how should the company balance revenue-generating projects with developing its own products and processes? It took time to discover that conventional catalysis (selling contract R&D services and systems) and proprietary bio-based products and processes would be the long-term focus of the company.

However, after the initial turbulence, we started to harvest business from our early investors. During those early years, the technical team worked on the core catalyst testing technology ("Flowrence"). The third-generation Flowrence technology could handle a much broader scope of chemical reactions and process conditions to accommodate customer catalysis projects. In addition, we managed to differentiate from our competitors with a rational and high-quality screening philosophy ("make every experiment count") and a straightforward business model (fee-for-service: all project IP would be assigned to the client, while we maintained IP on the equipment and testing methodologies).

Building on numerous feasibility studies to demonstrate the value and scalability of the Flowrence technology, we engaged in sizable R&D programs with leading refinery and chemical companies. This included a multiyear strategic collaboration with BP, which

How to Commercialize Chemical Technologies for a Sustainable Future, First Edition.
Edited by Timothy J. Clark and Andrew S. Pasternak.
© 2021 John Wiley & Sons Ltd. Published 2021 by John Wiley & Sons Ltd.

agreed to incorporate our automated parallel oxidation catalyst testing technology and modeling techniques in its world-leading purified terephthalic acid (PTA) business. This demonstrated our viability as a contract research service and systems company with a growing profitable business. Our venture capital shareholders were pleased with the positive financial turnaround and business results. They were faced with the choice to either sell the company or go for higher returns by pursuing higher-risk technology development opportunities.

14.2.2 Change in Direction

Our management team was convinced that the company could further leverage its catalytic process development platform. After careful evaluation of a few dozen opportunities, we initiated the development of a next-generation bio-based plastics and biofuels/fuel additives technology program. In 2006, bio-based chemicals were not as hot as they are today. The driver at that time was scarcity of raw materials (end-of-oil) rather than concerns about sustainability (CO_2 induced climate change). The industry focused on applying biotechnology and using (modified) enzymes or whole-cell technologies in fermentative processes. We took a different approach by setting out to use our catalytic process expertise and testing capabilities to convert sugars into a family of compounds called "furanics."

In 2004, the US Department of Energy issued an influential report called "Top Value-added Chemicals from Biomass" [1]. The 12 molecules with the greatest anticipated potential were evaluated with respect to economic and market potential. We extended the list and developed our own evaluation, realizing that production costs would be critical, especially for those chemicals already on the market as they are fundamentally the same molecules with identical performance regardless of their origins. For decades, the chemical industry has been narrowly focused on increasing production and reducing costs. This makes it extremely difficult to commercialize the same product from a different feedstock, as it will have to compete on cost from day one. Many sustainable chemistry companies have failed to recognize this challenge in time and had no plan on how to overcome this gap, leading to failure of the business.

It is different for new chemicals not yet on the market. These new molecules have the advantage of no fossil fuel analog to compete with. However, these new molecules will always be compared to existing comparable products. It's critical to investigate any advantageous properties they may have that bring value in specific product-market applications. There can also be a greater tolerance to price if you are solving an unmet need.

This brings us to the key challenge for the development of any novel chemical product and/or process technology. In the early days of commercialization, production volumes will be relatively small compared to incumbent, commodity chemicals. For example, the most modern terephthalic acid (TA) plants (the main monomer for PET plastic) produce more than 1 million tons per year. This brings an "economy of scale" that is impossible to compete with when building a first-of-a-kind commercial plant for a novel bio based product at an annual scale of 5000–25 000 tons. These relatively small plants still require multiple synthesis steps and require significant capital investments but cannot be skipped for the reasons detailed in Chapter 11.

To recap, any change or modification is much easier at a smaller scale. In addition, larger-scale plants come with larger investment requirements. When building a first-of-a-kind plant, the higher risk and investment has to be balanced with higher rewards. Finally, investors in such a plant require large-volume commitments to cover for the inherent risks of a new material entering the market. Ideally you would like to sell a significant fraction of the plant capacity to your customers even before the plant is built. Pre-selling volume in a nonexisting market is a challenge, especially when the production capacity is large.

So, why were we so interested to pursue furanics? As mentioned, production cost is the most important success criterion for bulk applications; feedstock cost is a big chunk of the overall production cost. For that reason, we looked at chemical conversions that at 100% yield retained most of the sugar mass in the target molecule. Furan dicarboxylic acid (FDCA) scores very high since we convert a C6 sugar into a C6 product. Assuming a 100% yield, 1000 kg of FDCA can be produced from 1150 kg of C6 sugar. This compares very favorably with TA, which in a 100% yield process would need at least 2170 kg of C6 sugar to produce 1000 kg TA. Bio-ethylene is even worse as 3215 kg C6 sugar is required to generate 1000 kg at 100% yield.

Furanics had been investigated for decades by industry leaders such as Quaker Oats, DuPont, and DSM. They were traditionally obtained by the acid-catalyzed dehydration of sugars and a catalyzed selective oxidation to FDCA to be used as a monomer for polyesters and nylons. Despite some attractive properties, furanics had not been commercialized because of the lack of a commercially viable production process for FDCA. This was exactly the opportunity we were looking for.

These two steps to convert C6 sugar to FDCA were a very good fit with our core expertise and infrastructure. For a century, researchers had developed sugar chemistry in water as one of the few good solvents for sugars. Leveraging our know-how, it was only a small step to change solvent for the sugar conversion step and address its solubility to unlock the door to a viable FDCA production process.

14.2.3 Exploring and Validating a New Opportunity

In 2005, when the plan was presented to the team and supervisory board, the general response was rather skeptical. People did not believe that a shift toward cleantech was a smart choice (oil was cheap, trading at less than $50 a barrel). Nevertheless, we decided to pursue the opportunity starting with a relatively small project team. We conducted catalyst screening experiments in combination with early conceptual process designs to develop economic catalytic processes. Once the full commercial potential of the new technology was better understood, we rapidly expanded the effort.

Looking back, though, while most people in the company realized the logic and importance of pursuing an internal product development program, it also created new challenges for us. First, it raised the question of how these programs were to be financed for the longer term. Second, it meant resource allocation conflicts between this internal program versus our established revenue generating customer R&D programs. The need for capital and the venture capital shareholder base pushing the company to create exit opportunities were driving management to another turbulent period.

14.2.4 Huge Challenges and Huge Advances

In 2007, we decided to pursue a listing at the Amsterdam Euronext Stock Exchange. We saw a rapid global increase in start-up companies pursuing biofuels and bio-based chemicals, with capital flowing into this new and exciting sector. In the background of our IPO preparations, the team had to ensure the company maintained momentum in its development activities and overall growth. In the spirit of our start-up company culture, our biofuels team carried out the first successful engine tests with furanics-based biofuels in a second-hand Citroen Berlingo. We hooked it up to analytical equipment to measure NOx, SOx, soot, and other engine parameters just two months before the start of the IPO roadshow. The timing of our stock listing turned out to be dreadful because of the collapse of the stock markets resulting from the US subprime crisis, and we ultimately had to withdraw our IPO the evening before going public. This was a major disappointment after the enormous time the team had invested in preparing the listing; but there was no time to regret, the company needed new sources of capital to continue our furanics work.

Some investors responded positively to the IPO road show, so the management team decided to attract new capital with a private financing round. In October 2008, four weeks after the dramatic collapse of the American investment bank Lehman Brothers, we successfully closed our Series E financing round of $26 million (€18 million). In 2009, our service and systems business was impacted by the economic recession as customers reduced external R&D spending. Despite this, we increased our investment to accelerate process and application development in FDCA-based chemicals and plastics. As part of this new strategy, we signed a collaboration agreement with Cargill subsidiary NatureWorks to develop FDCA-based polyesters. This partnership came at an excellent time.

We were producing FDCA at kilogram scale and wanted to pursue FDCA-based polyesters. NatureWorks was one of the few companies to commercialize a novel polymer in decades (polylactic acid [PLA]) and therefore understood the process of matching polymer properties with end-use requirements. We hypothesized that FDCA-based polyesters could be applied in areas where PLA had been unsuccessful, such as plastic beverage bottles. We produced a series of these polyesters, while NatureWorks evaluated their structural, thermal, mechanical, and barrier properties. One of these polyesters polyethylene 2,5-furandicarboxylate (PEF), turned out to be a standout candidate as it had excellent barrier properties along with enhanced thermal and mechanical properties compared to PET (see Figure 14.2.1). A strong barrier preventing oxygen entering the bottle (relevant for juice and beer) and CO_2 leaving the bottle (relevant for any carbonated beverage) was an unmet need in the market. The industry was using special coatings, additives (scavengers), or nylon layers for PET bottles to improve this barrier. This not only added cost but also caused a headache for the PET recyclers. PEF monolayer was a great step toward solving this unmet market need.

Luck and good timing can play an important role in innovation. At a climate convention in Copenhagen, a new PET plastic bottle partially made from plant-based sugars was launched. Brand owners using bottles for water and carbonated soft drinks (CSD) were looking for 100% plant-based materials as an alternative for their petroleum-based PET bottle. A meeting with one of these CSD brand owners was arranged during which we placed one of our first PEF bottles on the table. It still had a "golden" color (a consequence of insufficient monomer purity and initial polymerization conditions), but its list of properties sounded

Figure 14.2.1 Beneficial properties of polyethylene 2,5-furandicarboxylate (PEF).

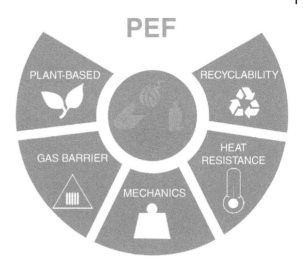

too good to be true. A material transfer agreement was signed allowing this company to independently evaluate PEF's properties. When its outstanding properties were confirmed, the first multiyear joint development agreement (JDA) was signed to further develop PEF for bottles. At the same time, a similar JDA was signed with Danone Waters (Evian) with a focus on water bottles and later ALPLA (a major PET converter from Austria) joined as a third partner. One of the first things that our partners requested was to stop calling this the "Furanics Technology" as ideally a new name should have no meaning (and certainly should not be anything chemical). We introduced the name YXY for our sugar-based FDCA and PEF technology.

The "bottle consortium" partners were valuable as they gave us access to regulatory expertise for obtaining food contact approval for PEF and toxicity testing of the FDCA monomer. ALPLA also owned PET recycling facilities and jointly performed sorting trials to demonstrate that PEF could be sorted out of a mixed plastic stream using widely applied infrared technology. The partnerships not only advanced the technical and regulatory development, but also resulted in a $50 million (€36 million) Series F raise involving all three partners alongside our existing shareholders. This important financing round allowed our company to complete two pilot plants for the key FDCA production steps (dehydration and oxidation) in the Netherlands. These pilot plants were vital for producing multiton quantities of FDCA and PEF required for application development, in addition to establishing a scalable process.

14.2.5 Expanding Our Technology Portfolio

This financing round also allowed the acceleration of a second technology that we were developing in the background. This technology targeted the other monomer required to make PEF (and PET): monoethylene glycol (MEG). As MEG is a drop-in, it was important to determine whether the technology at scale could compete on price with the low-cost

Figure 14.2.2   Avantium's monoethylene glycol (MEG) pilot plant.

incumbent route of naphtha to ethylene to ethylene oxide to MEG. Previous promising bio-MEG routes (four steps via bio-based ethylene) had proven to be too expensive for commodity bulk deployment. After determining that it could be cost competitive with the five-year average naphtha-based MEG production cost, we set out to develop a single-step route for producing MEG from sugars using a selective catalytic hydrogenolysis process. Meeting the required yield target under scalable process conditions led us to construct and commission a pilot plant in 2018–2019 to demonstrate what we called the RAY technology (see Figure 14.2.2).

14.2.6   Additional Strategies and Lessons Learned

Scaling a sustainable chemical technology is always more difficult than anticipated based on our experience with the YXY technology. When we started out in 2006, we envisioned developing a pilot process for FDCA to produce commercial quantities. Ten years later, we had spent more than $130 million (€100 million) validating the FDCA process at the pilot scale, along with purification to polymer-grade FDCA. We optimized the polymerization process to generate colorless PEF at molecular weights required for bottle, fiber, and film applications. But it did not end there. As we wanted PEF to be widely commercially available, we also invested in our own application development programs (bottles, fibers, and films), generating our own intellectual property positions. This was in addition to working with partners to test the most interesting applications of PEF for market introduction. We also investigated the recycling of PEF (both mechanical and chemical closed-loop recycling), its approval in food contact applications, and biodegradation trials.

It became clear that operating as a viable business at the pilot scale was not feasible and that we would have to demonstrate the technology at the commercial scale and have the product validated in the market. In 2015, an engineering team was established to work with an external engineering, procurement, and construction (EPC) contractor on a process design package for a 5000 ton per year FDCA commercial production plant. By contrast, PEF is produced for us by toll manufacturers as the assets required for this type of process (typically PET assets) at a range of scales are available globally.

At the start of the engineering phase, we also held discussions with scale-up providers to see whether partnering would be an attractive option. This would mean sharing costs and risk, but also gaining access to engineering capabilities and a sales organization to supply the product to the market. Of course, partnering also means sharing the upside. But sharing the upside of a much bigger pie can be an attractive way forward, especially if your business models are aligned, which is *not* a trivial starting point. In 2016, we opted for the partnering strategy, as we felt this would suit the needs of our downstream partners to use PEF in relatively large market applications, requiring certain economies of scale. BASF was selected as the best partner as we could leverage its expertise and build a 10-fold bigger plant (50 000 tons FDCA per year) together. BASF viewed FDCA and PEF as a compelling growth opportunity and not an obvious threat to its existing business.

Although BASF had to buy into our joint venture (JV) called Synvina, we would also financially contribute to the capital investment of this FDCA plant to be built on BASF's Antwerp site. To get access to this type of capital, our management team and supervisory board decided to list the company on the stock exchange to raise capital through an IPO. Following the establishment of the JV in late 2016, the IPO was completed in March 2017. The timing was much better this time, and investor interest exceeded expectations. We raised $120 million (€109 million) and secured our share for building the 50 000 ton per year FDCA flagship plant.

Unfortunately, our JV with BASF did not work out as planned. Together, the companies made strong technological progress through significant investments in the FDCA pilot plants to make the process ready for scale-up. However, we had differing views on how to best bring such a new monomer and polyester to the market. In early 2019, prior to construction of the plant, we acquired BASF's shares in the JV and by doing so regained full ownership of the YXY technology. In the following months, the company decided to pursue the original plan of building our 5000 ton per year FDCA plant.

Valuable time was lost to the JV with BASF, but a steep learning curve was successfully climbed. In mid-2019, a new plan was presented to the investor community. An estimated $175 million (€150 million) would be required to fund Avantium from 2019 to 2023, including the full amount for designing, building, and commissioning the first-in-the-world FDCA plant in the Netherlands. Worley was presented as our EPC partner and the front-end engineering and design (FEED) was kicked off in early 2020. At the time of writing, close to $115 million (€100 million) is secured, and we announced some of our strategic partners participating in the plant financing in January 2021. Construction should start in early 2021 and be complete in 2023. Eighteen years after conducting the first experiments to make furanics with sugar from our company restaurant, we hope to find PEF in our supermarkets.

14.2.7 Summary

When developing a new product, you can get lost in pursuing wild ideas and an endless list of possibilities. It is important to focus on application areas where the most value can be expected (unmet needs) and to work with partners that have deep application knowledge and access to customers. While it is tempting to be affiliated with major industry players, it is important to establish partnerships that are well balanced and most impactful for your business as described in Chapter 9. Most products in the chemical industry are first developed in high-value applications and often in niche markets, before finding use in high-volume markets (if ever).

Be ready for a marathon rather than a sprint. Nobody likes it when projects take more time and money to complete than anticipated, but it pays off to consider and actively work on scenarios for how to best cope with delays and unforeseen setbacks.

Finally, cash is king. Make sure you are well funded. A part of your team needs to be dedicated to raising capital. Once a financing round is completed, you need to pursue the next one. Most companies in the bio-based chemicals and biofuels space have been underfunded and as a result never reach their potential.

References

1 Werpy, T., Petersen, G., Aden, A., et al. (2004). Top value added chemicals from biomass. US Department of Energy. https://www.nrel.gov/docs/fy04osti/35523.pdf (accessed 4 November 2020).

14. Making an Impact: Sustainable Success Stories

14.3

Hazel Technologies

Counterintuitive Thoughts on Sustainable Entrepreneurship
Aidan R. Mouat

Hazel Technologies, Chicago, IL, USA

14.3.1   Blind Luck or Preparation?

The inception of a business is a noisy process. In signal processing, "noise" has two mean-
ings. The first is the intensity of a background relative to the intensity of a desired signal.
When data do not readily coalesce and many outcomes are observed within any dataset, the
data can be called "noisy." The second definition is signals that come at random and contain
no useful information, causing a given signal to appear less coherent than it truly is. Both
definitions apply to the process of entrepreneurship. It appears there is often no discernable
pattern of success or failure in the decision-making behind building an enterprise.

When looking at industry statistics, it becomes obvious that quick success is not the norm
for new enterprises. While quantitative data is typically limited to businesses financed by
venture capital, they still serve as a cautionary tale – on average, 68% of businesses fail to
raise a follow-up round in their first three fundraising stages (Seed, Series A, Series B).
Perhaps even more compelling, 81% of businesses fail to exit across all fundraising rounds
(https://medium.com/journal-of-empirical-entrepreneurship/dissecting-startup-failure-
by-stage-34bb70354a36).

The disproportionately high rate of failure in venture-backed companies typically breeds
one of two attitudes. The first attributes the noise to the influence of "luck" – a nebulous
quantity related to chance meetings, the discovery of key pieces of market information,
or the sum output of market pressures on the decision-making of high-value customers,
as examples. The second attitude, however, attempts to differentiate why a new business
should *not* be subject to these statistics – the "unfair competitive advantage" venture capitalists (VCs) seek to justify their investment.

Often, this perspective is summed up by the phrase "Fortune favors the prepared mind."
There are certainly elements of coincidence that cannot be ignored, but it is also a question
of what an entrepreneur makes of the opportunities presented to them. Pat Flynn, one of
the co-founders of Hazel Technologies, is fond of formulating this a different way: every big
problem in the world is an equally big opportunity for a business solution, so focusing on
identifying those problems naturally leads the business to where the opportunities lie.

How to Commercialize Chemical Technologies for a Sustainable Future, First Edition.
Edited by Timothy J. Clark and Andrew S. Pasternak.
© 2021 John Wiley & Sons Ltd. Published 2021 by John Wiley & Sons Ltd.

14.3.2 Hazel Technologies: How It Started and Where We Are Today

The story of how I came to co-found Hazel Technologies follows a similar trajectory. While pursuing my doctorate in chemistry in 2014, I was named a chemistry fellow for the Institute for Sustainability and Energy at Northwestern University (ISEN). This led to two outcomes.

First, I was given coursework that explored key issues in global sustainability. This was new territory for me, even though my graduate work focused on the use of chemistry to develop and support sustainable processes. I was designing and synthesizing heterogeneous catalysts for transforming renewable chemical feedstocks into platform and commodity chemicals. While this work was broadly housed under the umbrella of "sustainable chemistry," my focus lay far more on the technical detail of the bench-scale research and development (R&D) projects I undertook than on their applicability to scaled chemical engineering.

My ISEN coursework forced me to look at the world more broadly. When I did, I learned something. In most of our world's major verticals – energy, medicine, transportation, commerce, etc. – we have experienced a significant technological revolution over the last few decades leading to an entirely new norm of business. The "disruptors" of the market take advantage of paradigm shifts in technology and market attitude and, in doing so, develop new categories of market demand (editors' note: this concept is explored in Chapter 2). The average consumer thinks of a "taxi" and a "rideshare" as two different things, even though to that consumer a rideshare is just a taxi summoned through an app.

However, when we approached the subject of agriculture, I realized that we have not yet had that paradigm shift. The last transformative technological revolution in agriculture, in my opinion, was the discovery of the Haber-Bosch process for fixing nitrogen into ammonia in the early twentieth century. This chemical process enables the production of nitrogen-rich fertilizer for crops, and has impacted the world such that 50% of the global population is alive today because of the calories generated from this process.

We founded Hazel Technologies in 2015. That same year, the US Department of Agriculture issued its first-ever mandate that we must reduce food waste by 50% by 2030 to meet the goals of the Paris Climate Accord (https://www.usda.gov/oce/foodwaste/faqs.htm). The National Resource Defense Council published its famous "Wasted" report in 2012, indicating that nearly 40% of food produced goes to waste rather than being eaten (https://www.nrdc.org/sites/default/files/wasted-food-IP.pdf). According to the Food and Agriculture Organization, if food waste was a country, it would be the third largest emitter of greenhouse gases in the world (http://www.fao.org/3/a-bb144e.pdf). Extrapolating from these and other data points, I developed the theory that we have reached the end of the agricultural revolution the Haber-Bosch process initiated. The new "big problem" was not production, but over-production and under-utilization. Reducing food waste was going to be the new paradigm in agriculture, and whoever commercialized a solution stood to tackle a problem worth up to $2.5 trillion. I wanted to apply my chemical knowledge to the problem of food waste, since the last time agriculture was fully transformed it was through a process in my same field of heterogeneous catalysis.

The second outcome of my involvement with ISEN was my enrollment in an accelerator-style course at Northwestern called "NUVention: Energy," which focused on providing an entrepreneurial toolkit to students interested in building a business. Prior to 2014, I had never considered running a business – I earned a BS and MS at Emory University in 2009 doing synthetic organometallic chemistry, worked in the medical device industry for a few years, and then went to Northwestern for my PhD. In fact, I was so certain that I would not be an entrepreneur that I complained to the ISEN program manager about NUVention. I insisted it would not be helpful to me and would distract from my doctoral studies. This was clearly my moment of luck, not preparation – he would not let me drop the course.

It was in NUVention where I met the other co-founders of Hazel Technologies. Except Dr. Adam Preslar and myself, none of the five of us had previously met. At our first meeting in December 2014, we sat in a coffee shop in Evanston, Illinois, and I told the team: "I think we have two options. We could create this business just for coursework and get nothing out of it but a good grade. Or, we could take it seriously and see where it goes." Everyone agreed we should focus on the latter, so then and there we made the decision that if the company had legs, we would let it change our lives. We incorporated in March 2015, and at the time of writing, three of the five original co-founders still work at Hazel Technologies.

As of 2020, we have raised almost $18 million in private equity capital. We have earned nearly $1 million in grant financing, largely from the USDA. Our principle base of operations is in Chicago, with some satellite operations in California and the Pacific Northwest. We have a head count of approximately 35 employees, and we service more than 150 clients in 12 countries. In 2020, we treated 3 billion+ pounds of produce, preventing nearly 250 million pounds of produce waste and reducing climate emissions by nearly 200 000 metric tons of CO_2E. Our rapid growth, capital efficiency, and understanding of multiple markets has put us on the path to profitability by 2021 – only the sixth year of the company's existence.

14.3.3 Understanding What Our Business Really Is

At Hazel Technologies, we make food last long enough. We invented a new technology that allows us to engineer ecofriendly materials to store and time-release active ingredients into the storage atmosphere of perishable food. We can use these active ingredients to target specific problems in food shelf-life, like ethylene damage in produce or microbial load in meats. The technology is passive and integrates with standard commercial packaging (see Figure 14.3.1). Since we only treat the storage atmosphere itself, we leave no residues on food and add no new chemicals to the food supply chain. Our sales process is fully business-to-business (B2B), with our typical customers being food producers and suppliers, distributors, and retailers.

I believe it is inaccurate to say that the Hazel business model exists to prevent food waste. It is more accurate to say that our business model creates demand for value-added food. While the distinction may seem subtle, it is a central issue for a sustainability company. Sustainability itself is not a product, but a positive externality on a product that must have all the same features as any other viable market solution. Customers must not only want sustainable products but be willing to pay for them. To have traction, a product must be best-in-class, with unit economics that yield return on investment to the customer while also yielding a profit for the business selling it.

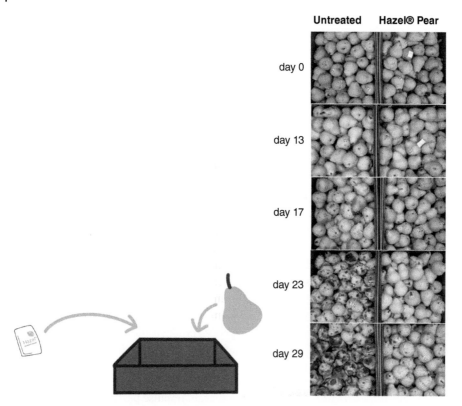

Figure 14.3.1   Pears stored in the absence and presence of Hazel Technologies sachet.

The concept of sustainability is challenging because it means different things to different people. Not all interpretations are correct. For example, it is popular to consider organic food more healthful than conventional food, though scientific studies indicate that there is no significant difference in caloric or nutrient intake between them. A consumer might think of something "non-GMO" as more sustainably produced, not realizing that non-GMO crops typically have higher susceptibility to disease, lower yield, and longer growing horizons than GMO equivalents, all of which impact the carbon and energy intensity of growing them. Conversely, a study from UC Santa Clara indicated that recent efforts to measure food waste have significantly underestimated the amount of pre-farmgate waste (i.e. field waste) that occurs before food is even put into the supply chain. To a grower, sustainability means being able to economically harvest and sell every piece of produce grown on their land, which requires access to enough markets to sell their crop and price points that a consumer will accept.

14.3.4   Targeting Value Through the Supply Chain

We approach the supply chain with economics top of mind. This includes obvious factors like the price of the product, which we factored into our product design from the beginning.

A sales guru taught me early in the process that there is no correlation between the price of a product and the cost of making it. While this is true, there are some chemical engineering ground rules that must be obeyed. We sought to reduce reliance on large, complicated packaging systems and incrementally improved packaging materials by using the packaging itself to deliver new biochemistry. In effect we set out to "functionalize" the storage environment. As a result, Hazel does not look like any other provider in the global market. We have no capital expense requirements, our products can be deployed by any unskilled operator, and we have strong margins at prices that are still accessible to our customers and therefore facilitate quick adoption.

It is also important to understand how the product creates value for the customer. Early on, we discovered that simply saving money on waste is not an interest driver for our customers. They are often concerned about adverse impact from labor issues, regulatory requirements, and cost of use. A more compelling proposition for them was the ability to increase market share by expanding the geographic radius of sale and being able to charge premium pricing for product. For example, in US tomato sales, about 85% of the revenue in the market is generated by about 25% of the total volume of tomatoes sold. Bulk pre-ripe green beefsteak tomatoes sell for $0.35/lb and have a large radius of sale. Ripe heirloom hothouse tomatoes sell for $4.99/lb but have a limited radius of sale. Our product allows the more expensive tomato to reach a larger customer base so the seller can generate more revenue from value-added versions of food.

It may seem odd to think of "value-added" food. Like any other product, consumers shop based on many strategies, like optimizing cost or maximizing wellness. Suppliers follow complementary strategies to maximize their margins. For example, grocery store design manipulates consumer choice to maximize high-margin impulse purchases. Furthermore, to be sold, food must gain value throughout the supply chain. Every hour of labor to pick and pack a piece of produce, every unit of energy spent refrigerating it, and every drop of gasoline that fuels its transportation becomes a part of the final price for the consumer. Hazel's shelf life enhancement technology is the same – the cost is factored into the overall technology stack of the food supply chain. If our product delivers more value than it costs, then we can reduce food waste by increasing the available demand for the food at the value-added price.

Understanding these factors not only enabled us to get quick market traction but were also critical to our success in securing venture capital. We started our Seed round for the business in mid-2016, at a time when many VCs were branching out of the "cleantech" space and trying to get into "agtech" without knowing much about it. Our biggest stumbling block was business model validation. We had the technical team, the customer interest, the intellectual property, and the unit economics on our side, but VCs lacked direct market experience in agriculture and would not believe in the business until customers started purchasing product. The influence of our market research on product design enabled us to move into the field quickly, gain those critical customers, and in turn successfully fundraise.

Our Seed raise did not close until February 2017. A significant part of the diligence process was direct customer visitation. Carsten Boers and Bernie Lupien, two of the principals of Rhapsody Venture Partners, went with us to customer facilities to watch our sales pitches and gauge customer acceptance of the product. It was a bit of culture shock for them, driving through hundreds of miles of agricultural area and meeting career farmers. In many ways that was the point – in the absence of scalable revenue metrics, there was no way for them to understand the relevance of our products to our potential customers without hearing from those customers how the product met the needs of their operations. The trip revealed key features of the product-market fit: for example, how our pricing fit into the operational expenses of a working packing house and how the small footprint and easy deployment of the product complemented hand-packing techniques in bulk specialty produce.

14.3.5 Final Thoughts

The sum of these market economics is critical to the inception and design of a business model. No entrepreneur ever has as much freedom in determining the success of their business as they do when the business is still in the earliest planning stages. If it does not work on paper, it certainly will not work in real life. The economic dimension is often the most neglected for first-time and student entrepreneurs, who may not see the fundamental link between the unit economics of their products and the opportunity for them to significantly impact the market. The emphasis in the phrase "Fortune favors the prepared mind" should not be on the role of *fortune*, but rather on what truly is a *prepared mind*. Venture capital tends to train entrepreneurs to think about scale before thinking about profitability; I do not believe one should be ignored in favor of the other.

As a first-time entrepreneur, some of the most meaningful advice I received was during a training session from an investor outside the venture capital space. He asked a group of us to articulate what kind of business we wanted to have. There are many kinds, and they do not all require the same strategy. A lifestyle business that provides a nice yearly salary for the family running it does not necessarily need to balance growth thinking with bottom-line management and profitability. In contrast, venture capital business accounts for less than 10% of all global commerce, but at accelerated growth rates that often lead to rapid market disruption. Venture capital businesses therefore think about growth trajectory and eventual profitability differently from most commercial enterprises. In any scenario, however, the entrepreneur who understands and analyzes all economic forces – customer priorities, the relationship of price and cost, the balance of near-term profit and long-term growth – stands the best chance of beating the average and creating a successful business.

Index

How to Commercialize Chemical Technologies for a Sustainable Future, First Edition.
Edited by Timothy J. Clark and Andrew S. Pasternak.
© 2021 John Wiley & Sons Ltd. Published 2021 by John Wiley & Sons Ltd.